Teacher Education for the Environment

European Perspectives

Teacher Education for the Environment

European Perspectives

Edited by

W. Leal Filho
and K. O'Loan

UNIVERSITY OF BRADFORD

MAKING KNOWLEDGE WORK

The Parthenon Publishing Group
International Publishers in Medicine, Science & Technology

NEW YORK LONDON

Published in the UK and Europe by
The Parthenon Publishing Group Limited
Casterton Hall, Carnforth,
Lancs LA6 2LA, UK

Published in the USA by
The Parthenon Publishing Group Inc.
One Blue Hill Plaza,
PO Box 1564, Pearl River,
New York 10965, USA

Co-published with ERTCEE
Department of Environmental Science
University of Bradford
West Yorkshire, BD7 1DP

British Library Cataloguing in Publication Data

Teacher education for the environment: European perspectives
1. Teachers – Training of – Europe 2. Environmental education – Europe
I. Filho, Walter Leal II. O'Loan, Kathryn
370.7'1'094

ISBN 1-85070-772-3

First published 1996

Typeset by European Research and Training Centre on Environmental Education

Printed and bound by Bookcraft (Bath) Ltd., Midsomer Norton, UK, on paper from a sustainable source

CONTENTS

LIST OF CONTRIBUTORS

Barbara Bajd
Pedagoska Fakulteta
University of Ljubljana
Slovenia

Simon Bailey
School of Education
Crewe and Alsager Faculty
Manchester Metropolitan University
Cheshire
England

Lídia Máximo-Esteves
Universidade do Minho
CEFOPE
Avenida Central 100
4700 Braga
Portugal

Walter Leal Filho
European Research and Training Centre (ERTCEE)
Abteilung Ökologie und Umweltbildung
Institut für Umweltwissenschaften
Universität Lüneburg
Wilschenbrucherweg 84
D-21335 Lüneburg
Germany

Chris Gayford
University of Reading
Department of Science and Technology Education
Bulmershe Court
Earley
Reading RG6 1HY
England

Sinikka Ojanen and Mauri Åhlberg
University of Joensuu
Savonlinna Department of Teacher Education
PO Box 55
FIN-57101 Savonlinna
Finland

Kathryn O'Loan
University of Bradford
ERTCEE
Bradford BD7 1DP
England

Klaus Schleicher
Universität Hamburg
Institut für Vergleichende Erziehungswissenschaft
Sedanstraße 19
D-20146 Hamburg
Germany

Tony Shallcross
Heriot Watt University
Moray House Institute
Edinburgh
Scotland

Arjen Wals, Kees Jongmans & Harm Biemans
Wageningen Agricultural University
Department of Agricultural Research
Hollandsweg 1
6706 KN Wageningen
The Netherlands

ACKNOWLEDGEMENT

We wish to extend a special note of thanks to Frances MacDermott, who assisted greatly with the publication of this book.

Walter Leal Filho
Kathryn O'Loan

INTRODUCTION

In order to be fully and successfully integrated into national education systems, environmental education needs to be considered not only in the context of curriculum policy, but also needs to be supported by class-based initiatives. For this reason the preparation of teachers, at both pre-service and in-service level, has a vital contribution to make.

This book was prepared by the European Research and Training Centre on Environmental Education (ERTCEE), as part of the Euro-Awareness project. Euro-Awareness is partly funded by Directorate-General XI (Unit B4) of the Commission of the European Communities. *Teacher Education for the Environment: European Perspectives* provides an overview of the variety of opinion as well as a description of a number of initiatives undertaken to strengthen the relationship between environmental education and teacher training.

The contributions included provide clear examples of approaches and methods that may be replicable elsewhere. It is expected that this book will be of use to teacher educators, co-ordinators, curriculum developers and researchers involved with or interested in environmental education.

CHAPTER 1

OVERVIEW

ENVIRONMENTAL EDUCATION AND TEACHER EDUCATION: SOME EUROPEAN PERSPECTIVES

Walter Leal Filho

INTRODUCTION

Much has been said about the subjects of teacher training and environmental education, but relatively little has in fact been done with respect to the critical examination of the relationship between both on an international scale, a line of thinking which also applies to Europe. The statements from the Centre for Educational Research and Innovation (CERI) on the interdisciplinary dimension of environmental education and Trent's (1976) record of the subject of teacher training as one of the important issues of the 1970s, show that twenty years down the road, the goal of integrating teacher training and environmental education remains as relevant as ever.

As stated by Leal Filho & Taylor (1995), it is widely acknowledged that environmental education involves not only the transmission of knowledge about the environment, a trend also pointed out by Fensham (1978), but the incorporation of cognitive and affective values that may stimulate people to change their attitude towards the environment and foster positive behaviour in relation to it.

Although on the one hand there are several ways in which environmental education may be incorporated into the context of a teaching system on an international scale, a practical question to be considered, on the other hand, is how to include the principles of environmental education in the teaching process in a way that caters for the specific needs of the formal educational process (Caduto, 1985). The traditional ways in which this has been conducted include:

(1) a subject;
(2) part of subjects;
(3) across the curriculum (cross-curricularly).

Environmental education "as a subject" is a strategy which foresees the establishing of a whole structure whereby a new subject or new discipline is created, with its own syllabus and work programme (Leal Filho, 1995). This is seen in part of countries such as Germany where the subject 'environmental education' is taught at some Universities (*e.g.* Hamburg) and in developing countries such as Guyana in South America (secondary level), although such a practice is widely seen as counterproductive in a developing country's context.

The advantage of this approach is that it enables attention to be focused fully upon environmental education and the opportunity for an in-depth analysis of the subject matter. On the other hand, the disadvantages of this approach include the fact that the focus needs to be fairly narrow to fit in the context of one discipline. In addition, there is a need to find space in the curriculum to accommodate yet again another subject. Moreover, the setting-up of a subject such as environmental education assumes that there are trained teachers, as well as supporting materials which will enable its implementation. There are few European countries where environmental education could be established through a discipline heading (Leal Filho, 1995).

Environmental education "as part of subjects" entails the incorporation of an environmental education component as part of individual disciplines, also requiring a full work programme. This is seen in European countries such as Portugal or Cyprus. In the latter, a 'project approach' to the dissemination of environmental education in formal teaching is used. The advantages of this approach are that it facilitates the dissemination of environmental education among a set of subjects; it caters for the one-by-one inclusion of environmental education within the framework of the programme of individual disciplines and facilitates debate.

The main disadvantage of this approach is that it traditionally restricts the dissemination of information on principles and practices of environmental education to those subjects which have a stronger link with the environment (*e.g.* science, geography), leaving others virtually uncovered (*e.g.* history, religion).

With regard to the dissemination of "environmental education across the curriculum", this approach attempts to ensure that environmental education is spread across the board (Keiny & Zoller, 1991), reaching all subjects, independent of their field of concentration (*e.g.* science, arts). This was the structure adopted in England and Wales, for example, over the past five years, despite some changes in the first half of 1995.

The main advantage of a cross-curricular approach is related to the fact that virtually all subjects have the opportunity to include environmental education in their programmes. One of the disadvantages of this approach is that it requires a great deal of organisation of the teaching system, with a view to enabling the dissemination of information to occur evenly, rather than being concentrated on specific subjects. In addition, as the example of England and Wales illustrates, such a modality is vulnerable to political changes and, whenever there are budgetary cuts and threats of reducing expenses, environmental education is one of the first victims. Other ways to introduce environmental education in the curriculum of European nations have also been recorded. In Malta, for example, an 'infusion' approach has been tried (Ventura & Pace, 1994), although it was seen that albeit feasible, such a strategy was not always easy, especially with regard to the exam subjects. Moreover, ideas aimed at guiding the incorporation of environmental education in the curriculum have been amassed by (Hungerford & Peyton, 1986).

A common feature of the above approaches to environmental education is that they require a number of back-up elements, especially:

(1) Reading materials and written references for teachers and pupils;

(2) Support and guidance for the wide spread handling of environmental education in a way that seeks to control duplications and shortages;

(3) Provisions for the training of teachers;

(4) Suitable time in the school curriculum and adequate time-table arrangements.

The training of teachers for the handling of environmental education is seen as an important part of the process. Although one cannot seriously regard it as the 'priority of the priorities' due to the equal relevance of the other items, it needs to be acknowledged that the provision of education for teachers, both at the pre-service or at the in-service stage, is very important for the implementation of environmental education in formal teaching.

The subsequent parts of this chapter will outline the benefits of the provision of environmental education components as part of teacher education programmes, the most common problems encountered and some measures that might be helpful in addressing the difficulties seen today on a European scale.

ENVIRONMENTAL TEACHER EDUCATION: EXPLORING THE LINKS

In attempts to analyse environmental education and teacher education in Europe, a description of the links between both items would seem to be a necessary procedure in identifying how best they may be integrated. At the outset, it should be recognised that environmental education needs to compete for time and space in the curriculum with many subjects and with a series of interdisciplinary topics, such as:

(1) health education;
(2) sex education.

These two areas of knowledge, used as simple examples, which can be regarded as well established and widely implemented in formal education, rely upon a substantial amount of resource material. They are also highly regarded by influential government agencies such as those in charge of education (*e.g.* Ministries of Education) and health (Ministries of Health). It is against this background that environmental education should be viewed.

Throughout Europe, there are various institutions concerned with the provision of in-service teacher training in environmental education. Table 1.1 provides a general overview of some organisations working within Europe to provide in-service teacher training. Although the given list is not meant to be comprehensive, it shows that various government and non-government agencies are working in this field. It should, at this stage, be stated that, albeit fairly wide spread, the provision for environmental education in teacher education in European nations is not even. There are on the one hand those countries where such a task is performed by a wide range of institutions, while on the other hand, there are countries where a few organisations (governmental or non-governmental) are left with the task of training teachers on environmental education.

As stated by Leal Filho (1995), because of its interdisciplinary nature, environmental education may included in the domain of:

- social sciences;
- natural or life sciences;
- exact sciences.

The possibilities for applying environmental education to teacher training, both at the pre-service and in-service levels, are therefore many. The incorporation of an environmental education perspective into the broad context of teacher training programmes may lead to the following developments:

Table 1.1 Overview of some institutions providing in-service environmental education training for teachers throughout Europe

Country	*Institution*	*Service Provided*
Austria	ARGE-Umwelterziehung	A, B, C
Belgium	Réseau IDée	A, C
Denmark	Royal Danish School of Educational Studies	A, B
Finland	University of Joensuu	A, C
France	Réseau Ecole et Nature	A, B, C
Germany	Teacher Training Institutes (Instit. für Lehrfortbildung)	A, B, C
Greece	Hellenic Society for Nature Conservation	A, B, C
Ireland	Ministry of Education	A, C
Italy	Regional Institute for Research, Experimentation and Educational Development	A, B, C
Malta	University of Malta	A, C
Netherlands	National Reference Centre for the Environment	A, B
Norway	Ministry of Education	A, C
Portugal	Portuguese Association for Environmental Education	A, C
Scotland	Scottish Environmental Education Council	A, C
Spain	National Centre for Environmental Education	A, B, C
Sweden	National Agency for Education	A, B, C

Key: A = organisation of training courses, seminars, workshops
B = preparation of teaching materials (*e.g.* books, leaflets)
C = support to school initiatives/provision of advice

(1) the examination of the major environmental issues from the local, national, regional and international points of view, so that pupils receive insights into environmental conditions in other geographical areas and in the place they live;

(2) enable learners to have a role in planning their learning experiences and provide an opportunity for making decisions and accepting their consequences;

(3) relate environmental sensitivity, knowledge, problem-solving skills and values clarification to every age, but with special emphasis on environmental sensitivity to the learners' own community in early years;

(4) help learners discover the symptoms and real causes of environmental problems;

(5) emphasise the complexity of environmental problems and thus the need to develop critical thinking and problem-solving skills;

(6) use diverse learning environments and a broad array of educational approaches to teaching / learning about and from the environment with due stress on practical activities and first-hand experience.

Attempts to address the above items will be reflected in the preparation of pupils in order to understand the complexity of the environment in which they live. The need for the comprehensive preparation of future teachers, as well as those currently in practice, to enable them to deal with environmental issues and foster pupils' environmental awareness is also an implication to be considered.

ADDRESSING THE PROBLEMS

International experience has shown that there are three main barriers to the systematic training of teachers in environmental education. These are:

(1) ***the provisions for training:*** excepting the pre-service training provided by universities and other educational institutions (*e.g.* Teaching Colleges), teacher training on environmental education is often sporadic. This greatly contrasts with training in areas such as environmental management or environmental pollution, topics on which, as described by Williamson (1995) training is fairly widely available all year round.

Training programmes on environmental education other than those undertaken by government agencies in charge of it, tend to occur fairly infrequently, are usually undertaken over short periods of time - as opposed to longer, more substantial courses or seminars - and in times of the year when the likelihood of participation in not very high (*e.g.* summer months, weekends).

(2) ***the duration of the training:*** as stated above, short-term training is, due to a combination of logistical and financial reasons, more favoured when compared to initiatives which take more time. The provision of such a modality of 'snap shot' training has the disadvantage of preventing a teacher on a pre-service or in-service basis, to fully acquire or comprehensively develop the skills needed to systematically

approach environmental matters with pupils once they complete the training programme.

(3) ***the nature of training:*** people have different levels of ability and different levels of information on what environmental education in fact is. The nature of some current training initiatives may enable a teacher to become familiar with some aspects (*e.g.* environmental education applied to sustainable development is a popular theme at the moment), but not with others. In such circumstances, some of those attending training programmes feel that their training and indeed their information needs are not being met and that they are not obtaining the help they need to provide a meaningful contribution to the environmental education of pupils.

Various attempts to address these problems have been made. UNESCO-UNEP's International Environmental Education Programme has, for example, published various environmental education modules suitable for use in a variety of contexts. These publications, available as part of the "Environmental Education Series", have been helpful in providing an overview of the various elements that integrate teacher training on the environment. Some titles relevant to teacher training are:

- Environmental Education Module for Pre-Service Training of Teachers and Supervisors for Primary Schools (UNESCO-UNEP, 1983a).
- Environmental Education Module for Pre-Service Training of Science Teachers and Supervisors for Secondary Schools (UNESCO-UNEP, 1983b).
- Environmental Education Module for In-Service Training of Teachers and Supervisors for Primary Schools (UNESCO-UNEP, 1985a).
- Environmental Education: Principles of Teaching and Learning (UNESCO-UNEP, 1985b).
- Interdisciplinary Approaches in Environmental Education (UNESCO-UNEP, 1985c).

Although not essentially aimed at dealing with European issues, the above publications provide some useful insights into the various procedures that need to be implemented to incorporate environmental education in the context of the training of teachers.

On a European scale, various attempts to promote environmental education in the framework of teacher training have been seen. For example, the Nordic Council of Ministers is supporting an environmental education project which caters for the training needs of teachers. In addition, the Association of Teachers Educators in Europe (ATEE) set-up a working group on environmental education which has been active in discussing various approaches to further environmental education as part of teacher training. Moreover, the European Teacher Education Network (ETEN), undertaken in the context of the ERASMUS programme of the Commission of the European Communities, has been trying to establish mechanisms to promote teacher training initiatives in the field of environmental education. One of the examples of such schemes is the joint preparation of a module on environmental education to pre-service primary school teachers between the Universities of Kalmar (Sweden) and Derby (UK) (SGEEE's Newsletter 3, 1995/6).

Recent events have also addressed the value of teacher training in environmental education. For example, the "European Colloquium on Environmental Education" held in Bradford, UK in February 1994, discussed the matter of teacher training is this field, recognising it as an important tool in fostering environmental education in Europe (Leal Filho, 1994), a trend supported by the European Seminar on Environmental Education Research, held in March 1995, which presents a research agenda for environmental education in Europe (Leal Filho & MacDermott, 1995). An initiative of the Directorate-General XI, Unit B4 (Environmental Education) of the Commission of the European Communities, under a 'Summer School' format, held in the summer of 1994 in the city of Toulouse, France, also addressed the issue of teacher training, which was perceived as an important component of attempts to promote environmental education in Europe. Publications such as "Practices in Environmental Education in Europe" (Leal Filho, MacDermott & Murphy, 1995) and "Environmental Education into Initial Teacher Education in Europe" (Brinkman & Scott, 1994) have also documented experiences in teacher training with a focus on environmental education.

The above state of affairs, combined with one reality, namely the different levels of priorities afforded by European countries to environmental education, may be indicative of the fact that pan-European approaches may be useful tools in enhancing the quality of teacher training on environmental education in Europe. A European wide training programme, training module or training initiative on environmental education needs to be:

(1) Comprehensive enough to meet an individual's information needs;

(2) Flexible enough to be carried out in the right time, either fully or in part, so that time-constraints do not preclude participation;

(3) Able to allow the enrolment of those who missed previous opportunities on a continuous basis;

(4) Oriented towards catering for the language differences and limitations of the audience.

In addition, there are also some fundamental principles in the provision of teacher training on environmental education that need to be considered (Figure 1.1).

The preparation of a European module on environmental education such as MESAGES (aimed at in-service training at the secondary level), described elsewhere in this book, has provided an important contribution towards addressing the above items. Further initiatives are needed, especially programmes which may be helpful in the pre-service phase. The preparation of an Europewide environmental education module for use in pre-service training, both for the primary and secondary level, is a challenge that, if met, may provide a significant step ahead in terms of integrating teacher training and environmental education and in ensuring that a European dimension is part of the process.

CONCLUSIONS

Despite the perceived importance of teacher training in the development of environmental education, there is a diversified and uneven provision of initiatives in this field in Europe today. Part of the problem is found in the way different countries treat environmental education and the varied level of support it receives. It is believed that the current amount of support for environmental education, acknowledged as unsatisfactory in some contexts, could be significantly enhanced, especially in those countries where it does not enjoy a high political degree of support, through the provision of a Europewide module aimed at supporting environmental education training at the pre-service primary and secondary levels. The experience gathered through the preparation of the MESAGES, an in-service environmental education module for the secondary level, substantiates the claim that provisions for Europewide teacher training on environmental education can and should be supported by appropriate tools and that it needs to be technically sound and flexible enough to be realistically implemented.

Such an approach would be in line with the true spirit of European co-operation and in line with the need to use techniques and methods that can integrate countries and ensure that environmental education is afforded the degree of emphasis it is entitled to and indeed needs, in order to fulfil its role as a tool for sustainable living and as an instrument for sustainable development.

Figure 1.1 Fundamental principles of teacher training in environmental education

Providing information

Training programmes on environmental education need to offer teachers information from a wide range of perspectives. This includes information on the definition, evolution, principles and practices of environmental education, as well as its links with biodiversity conservation and the ultimate goal of sustainable development.

Facilitating innovation

Training schemes need to facilitate the development of new approaches and the dissemination of teaching methods that reflect state-of-the-art information on environmental education to a practical context. Research, development and demonstration need to be more used as part of training initiatives on environmental education.

Encouraging new approaches

Through a variety of new teaching approaches and methods, and by using well-established ones such as role-play, simulation and gaming, teachers need to be able to raise motivation among pupils on environmental issues and to develop a commitment towards being involved in the process.

Engaging pupils

Training in environmental education should move on from the passive acknowledgement of the role of the 'head', 'heart', 'hands' approach to consider it as an extra tool towards engaging pupils in the environmental conservation process. Attempts to involve pupils, which have a proven impact in the development of their environmental awareness should be a prominent part of teacher training schemes.

Learning and working together

In the past, training initiatives in environmental education have paid some emphasis to the contribution of pupils to the learning process, but this needs to be increased. Training should show to teachers the real benefit of gathering and providing due value to the inputs of pupils, ensuring that their experiences are taken into account.

REFERENCES

Brinkman, F. G. & Scott, W. A. 1994. *Environmental Education into Initial Teacher Education in Europe* Brussels: ATEE.

Caudto, M. J. 1985. *A Guide to Environmental Values Education* Paris: UNESCO.

Centre for Environmental Research and Innovation (CERI) 1972. *Interdisciplinarity and Environmental Education* Paris: CERI.

Fensham, P. J. 1978. Stockholm to Tbilisi - the Evolution of Environmental Education. *Prospects 8*, pp 446-455.

Hungerford, H. R. & Peyton, R. B. 1986. Procedures for Developing an Environmental Education Curriculum. *UNESCO-UNEP International Environmental Education Programme, Environmental Education Series no. 22*. Paris: UNESCO.

Keiny, S. & Zoller, U. (Eds.) 1991. *Conceptual Issues in Environmental Education* New York: Peter Lang.

Leal Filho, W. (Ed.) 1994. *Trends in Environmental Education in Europe. Proceedings of the 'European Colloquium on Environmental Education', University of Bradford, UK, February 1994* Bradford: European Research and Training Centre on Environmental Education.

Leal Filho, W. & MacDermott, F. 1995. *Proceedings of the European Seminar on Environmental Education Research.* Economic and Social Research Council, Commission of the European Communities and ERTCEE. Bradford: European Research and Training Centre on Environmental Education.

Leal Filho, W. MacDermott, F. & Murphy, Z. 1995. *Practices in Environmental Education in Europe*. Bradford: European Research and Training Centre on Environmental Education.

Leal Filho, W. 1995. The integration of environmental education in the context of curriculum systems in an international context. In: Proceedings of the IRSSAE Conference. Florence, Italy, 11-13 September 1995.

Leal Filho, W. & Taylor, C. 1995. The Integration of Environmental Education Into Teaching Systems. In: ASESP Seminar on Environmental Education. Arusha, Tanzania, September 1995.

Study Group on Environmental Education in Europe 1995. *SGEEE Newsletter, Issue 3, Winter 1995/6*. European Research and Training Centre on Environmental Education.

Trent, J. H. 1976. Changes and Trends in Environmental Education in 1972-1975. *Journal of Environmental Education 7* (31), pp 51-60.

UNESCO-UNEP International Environmental Education Programme 1983a. Environmental Education Module for Pre-Service Training of Teachers and Supervisors for Primary Schools. *Environmental Education Series No. 5.* Paris: UNESCO.

UNESCO-UNEP International Environmental Education Programme 1983b. Environmental Education Module for Pre-Service Training of Science Teachers and Supervisors for Secondary Schools. *Environmental Education Series No. 8.* Paris: UNESCO.

UNESCO-UNEP International Environmental Education Programme 1985a. Environmental Education Module for In-Service Training of Teachers and Supervisors for Primary Schools. *Environmental Education Series No. 6.* Paris: UNESCO.

UNESCO-UNEP International Environmental Education Programme 1985b. Environmental Education: Principles of Teaching and Learning. *Environmental Education Series No. 20.* Paris: UNESCO.

UNESCO-UNEP International Environmental Education Programme 1985c. Interdisciplinary Approaches in Environmental Education. *Environmental Education Series No. 14.* Paris: UNESCO.

Ventura, F. & Pace, P. 1994. Environmental education across the curriculum. In: Farrugia, C. (Ed.) *A New Vision for Primary Schools.* Valetta, Malta: Malta Union of Teachers.

Williamson, C. R. 1995. Training in Environmental Education. Unpublished MPhil thesis. University of Bradford.

CHAPTER 2

ENVIRONMENTAL EDUCATION, TEACHER IN-SERVICE TRAINING AND SCHOOL DEVELOPMENT IN THE NETHERLANDS

Arjen Wals, Kees Jongmans & Harm Biemans

INTRODUCTION

In this contribution we discuss some current developments in teacher in-service training and school-development in the Netherlands and their relationship with environmental education. Dutch secondary schools as well as vocational schools are presently undergoing significant changes in the fields of school organisation and curriculum. Schools are becoming more and more autonomous in the way they organise their curriculum, teacher in-service training and school management. Many teachers and school managers support the proposed changes, but find it hard to bridge the gap between educational theory and practice. Environmental education provides many examples of how to implement some of these changes. Our main presupposition is that environmental education can act as a catalyst for educational change or reform towards participatory models of teaching, learning and school development. The goals of environmental education, as we see them, as well as the process of environmental education seem to match the new educational policies which are to lead Dutch education into the next century.

ORGANISATIONAL CHANGES IN DUTCH EDUCATION

To start, let us briefly look at an important change occurring in the Netherlands at the level of school organisation: the creation of more autonomous schools. The national government now provides schools with a budget of their own with less strings attached than before. Schools are becoming more responsible for shaping their own identity and creating high quality education. In essence school organisation and development is becoming more market-driven. For instance, teacher in-service training institutes now have to compete with each other for a school's in-service training budget, while schools now compete with each other for student enrolment. The general idea behind decentralised education is that less government involvement will reduce cost and at the same time will increase the efficiency and quality of education.

At the level of the curriculum, teaching and learning the Ministry of Education, Culture and Sciences is now advocating a shift from teacher-centred teaching to student-centred teaching. This shift is accompanied by an emphasis on the development of abilities and skills, as opposed to an emphasis on the transfer of disciplinary knowledge. These abilities and skills include: problem-solving, working in groups, action taking, conflict management, critical thinking and learning to learn. The latter skill, learning to learn, could be regarded as a key skill in contemporary education for it allows students to develop meta-strategies in dealing with often conflicting and overwhelming information in the context of problem-based learning. However, the government at the same time provides national standards for each subject taught in schools, in essence, still suggesting a pre-occupation with the transfer of disciplinary knowledge. As students need to pass national exams at the end of secondary school and schools are evaluated on the percentage of students graduating, teachers are often not inclined to deviate from the traditional curriculum.

Obviously, these proposed shifts at the school organisation level and at the curriculum level require a rethinking of the role of teachers and students in the learning process. A teacher is to assume the role of a facilitator who, depending on the situation, provides information, counsels and monitors individual students, coaches the class, helps students monitor themselves, distils meta-cognitive strategies, helps students become confident in what they do, and, finally, frequently becomes a co-learner with the students. The students, on the other hand, will have to assume the role of an active participant who works with other students, utilises a variety of information sources, contributes to the problem-solving process, shares his or her own experiences, monitors his or her own learning, and frequently becomes a co-teacher with the teacher when sharing personally gathered information with others in the classroom or in public (see also: Resnick, 1989; Simons, 1991; Vermunt, 1992; Wals *et al.*, 1990; Stapp & Wals, 1994).

ENVIRONMENTAL EDUCATION AS A CATALYST FOR EDUCATIONAL CHANGE

In our view environmental education can be broadly defined as the process that enables students and teachers to participate in the planning, implementing, criticising, and evaluating of educational activities aimed at resolving an environmental issue that they themselves have identified (Wals *et al.*, 1990; Wals, 1994a, 1994b; Stapp & Wals, 1994). What an "environmental issue" is, then, depends on the perceptions and earlier experiences of the learner as well as the context in which education takes place. Environmental problem solving is seen as the process that is needed to resolve an environmental issue. In other words, environmental education should be a learning process that seeks to enable participants to construct, transform, critique, and emancipate their world in an existential way. *Construct* in the sense of building upon the prior knowledge, experiences and ideas of the learner. *Transform* in the sense of changing, shaping, influencing the world around them, regardless of scope or scale. *Critique* in the sense of investigating underlying values, assumptions, world views, morals, etc., since they are a part of the world around the learner and since they are a part of the learner him/herself. *Emancipate* in the sense of detecting, exposing and, where possible, altering power distortions that impede communication and change. Personal experience and involvement are two essential driving forces of environmental learning processes.

The above ingredients are also essential for realising educational change while the 'definition' of environmental education as provided here includes several elements that can also be found in recent government documents on changing Dutch secondary and vocational schools (i.e. Procesmanagement Basisvorming, 1993). Hence, in order to broaden and deepen environmental education in Dutch secondary and vocational schools, environmental educators will have to show how environmental education can easily become the vehicle on which education change can take place.

PROFESSIONAL DEVELOPMENT AND TEACHER IN-SERVICE TRAINING

What are some of the conditions under which educational change can blossom? When regarding environmental education as educational innovation, we have to consider the conditions a school should meet in order to truly innovate. Some research has indicated that the extent to which schools are able to change largely depends on their ability to develop their own policies, strategies and guidelines for educational and organisational change. A school as an organisation is more able to shape its own future in a democratic way when teachers are an integral part of any effort to develop the above mentioned policies, strategies and guidelines (Sleegers, 1991). Unfortunately, teacher involvement in the development of school policies, strategies and guidelines is often lacking (Jongmans, 1995). Jongmans (1995) also shows that a teacher's involvement in school policy-making depends on his or her own perception of his or her role as a teacher.

Research has shown that teachers have different professional orientations. Hoyle (1980) distinguishes two professional orientations: a restricted professional orientation and an extended professional orientation. Restricted professionals can be characterised as teachers who consider teaching (both its process and its contents) to be the responsibility of the individual teacher. Extended professionals, on the other hand, can be characterised as teachers who regard teaching as a predominantly social activity in that fellow teachers, school managers, teacher trainers and educational researchers have a shared responsibility in optimising teaching and learning. Based on an extensive study conducted in Dutch agricultural schools at the secondary level, Jongmans (1995) distinguishes a third type of professional orientation, namely, the less extended professional. He suggests that, instead of speaking of a dichotomy, one can speak of a professional orientation continuum. A less extended professional does recognise the importance of discussions with fellow teachers on the subject of teaching and learning, but regards the selection of learning activities and learning contents to be the sole responsibility of the individual teacher. Furthermore, the less extended professional is less inclined to compare his or her own teaching methods with those of others, is less involved in school policy-making and is less motivated to participate in teacher in-service training.

When considering environmental education as both an educational innovation on its own merit and as a means to facilitate educational change, environmental education in-service training cannot ignore a school's policy-making capacity and the professional orientation of teachers. Teacher in-service training in environmental education should not only focus on didactical methods and learning contents, but also on empowering and equipping teachers to re-design the school curriculum and to contribute to school policy-making. Prior to developing any environmental education in-service training a school should chart the professional orientation continuum of its staff. Not until then can a school try to move its overall professional development orientation from restricted to less extended to extended. Schools that employ a relatively large group of teachers who have developed an extended professional orientation and school managers who favour participatory models of policy-making, are more willing and able to respond to desired changes in teaching, learning and curriculum. Clearly, environmental education teacher in-service training and professional development should be mutually reinforcing.

In the introduction we mentioned the increasing autonomy of Dutch schools in the way they organise their curriculum, teacher in-service training and school management. For example, since 1993 schools are expected to develop and implement their own teacher in-service training and professional development programmes and strategies. The government now provides schools with their own budget for teacher in-service training. This autonomy is at least partly driven by unsatisfactory results of the old top down or supply-driven approach to teacher in-service training and professional development. Prior to 1993 teacher in-service training institutes received a budget for teacher in-service training and professional development of school teachers. The programmes these institutes offered often did not match the demands of schools. The idea is that when schools are responsible for their own teacher in-service training and professional development, they are more likely to address the immediate and specific needs of the school as a whole and of individual teachers (see also Jongmans, 1994a). The Dutch government expects the effectiveness of teacher in-service training and professional development to increase in the new more school-centred model.

Studies by Jongmans (1994a, b; 1995) show, however, that this increase in effectiveness, at least in the case of agricultural secondary schools, has not yet been realised. Apparently schools neither have a tradition in assessing their own school and teacher specific needs nor in developing an effective teacher in-service training and professional development programme. In addition it turns out that the school policy-making capacity is rather limited at this point. Under these circumstances schools will find it hard to support and develop a professional development programme for environmental education. What we are currently seeking to cultivate is a strategy for the implementation and development of environmental education which integrates aspects of teacher in-service training, professional development and school policy-making.

SOME GUIDELINES FOR TEACHER IN-SERVICE TRAINING

One of our main presuppositions is that if a school wants to realise educational change then it is crucial that all parties involved jointly assess and define the situation the school is in, how it arrived at that situation, the imagined outcome of the educational change process and the route that could transform the present situation into a more desirable one (Robottom, 1987; Elliott, 1991; Wals, 1993, 1994c). This view of school-based development, implementation and evaluation differs from a more traditional view of educational change which holds that outside experts can best diagnose schools and develop prescriptions that could lead to the necessary changes. Particularly in situations where teachers feel that they are rarely consulted, the model of expert-based innovation no longer seems appropriate.

Given the importance of involving teachers in the development, implementation and evaluation of educational change or an educational innovation, an action-research approach appears most appropriate for developing environmental education in schools. Action-research here means the blending of environmental education theory and environmental education practice in such a way that those who will be most affected by any changes that result from the research are also the ones who help determine what theories and experiences are the most meaningful and relevant. The blending of theory and practise involves *praxis*, a spirally developing concurrent process of evaluation, reflection and action (Lewin, 1946). While participants are engaged in a problem-solving process, they are continuously reflecting on their work and its effectiveness. This is an active process in that the participants seek out the information they need, define their own problems and develop a range of solutions.

During action research the participants can be assisted by an outside facilitator whose task it is to create the right conditions for action-research, to facilitate praxis, to bring in new information and to systematically collect the participants' experiences in order to document the research and disseminate the findings to other settings. The facilitator, who could be working for a teacher in-service training institute, not only helps the participants in their own inquiries by bringing in relevant educational theory or didactic methods, but also benefits from participation in the process him or herself (Stapp & Wals, 1994; Wals, 1993; 1994b). It is through direct interaction with the participants in the school and by 'being there' that the facilitator is able to get a better sense of what goes on in the schools. This 'sense' is absolutely essential for tailoring a teacher in-service training and professional development programme to the participating school.

Based upon our own prior research we have designed several guidelines for the advancement of teacher in-service training and professional development in environmental education (Alblas, *et al.*, 1995; Biemans & Simons, 1992; 1995; Jongmans, 1994a; 1995; Alblas *et al.*, in press):

School-based

Regardless of whether environmental education is integrated in the curriculum, a separate course or a module, it is crucial that it is rooted in school-wide support. This will require involvement of teaching and management staff in developing a sense of purpose for including environmental education in the school's array of activities and in the design of the curriculum. Although models and examples from elsewhere may be useful in building images of what could be done, they can never be recipes for implementation.

At the onset of developing an environmental education teacher training programme a school will need to analyse the current state of environmental education in both the school and in the curriculum from the perspective of school management, teachers and students. School management will have to express its own needs and wishes in terms of goals of environmental education and the type of school organisation and policy that could help realise these goals and wishes. Likewise teachers will have to express their needs and wishes, relate them to their perspectives on teaching and learning (Hoyle, 1980), determine ways to contribute to the development of school organisation and policy suitable for the desired environmental education, assess their own teacher in-service training and professional development need in this area, and, finally, critically look at existing teaching materials and expected learning outcomes (national standards). Students will have to be consulted on the way they perceive the teaching of their teachers (see also Rosenshine & Stevens, 1986; Van Amelsvoort *et al*, 1993), the way they themselves learn (see also Riteco-Slaats, 1994; Vermunt, 1992), the way they perceive the various subjects taught in school (Bergen *et al*, 1994; Kuhlemeier *et al*, 1990), and, finally, the knowledge, insights, attitudes and skills they have already developed with relevance to environmental education (Wals, 1994b).

Purposeful

Teachers and other staff members will have to reflect on the goals and objectives of their environmental education programme from a variety of vantage points. What is in the best interest of the learner? What is in the best interest of the school and the community? What's in the best interest of society as a whole? What's in the best interest of the planet? A discussion on environmental education goals and objectives could help teachers and the school as a whole to develop a sense of purpose, motivation and commitment regarding developing the curriculum. Such reflections and discussions should be on-going.

Learner centred

In order to involve the learner in environmental problem solving, teachers will have to get to know their students. What are their concerns, interests, expectations and ambitions in general, but particularly in the light of the environmental crisis? What (mis)conceptions have they developed to understand or explain environmental phenomena? How have these concepts been formed and how do they relate to those used by others? Teachers should take ample time to elicit students' ideas and feelings and make them subject of classroom debate. Not only would it help teachers in adapting the learning process and contents to the individual needs of students, it could also motivate students by letting them know that their voices are important.

Contextual

Each community, school and individual is different. Consequently duplicating a lesson or a series of learning activities for use in a different community, school or even a different class, does not do justice to the uniqueness of each situation. Some adaptation to changing or varying circumstances is required to make an environmental education programme meaningful for the learner. The contextualisation of the curriculum and its programme of activities does not free the teacher from making links to situations elsewhere. When possible teachers should spirally move beyond local community issues to similar issues on a larger scale (regional, national, global), but always using the local community as a frame of reference. By doing so, the learner is able to see that the local environmental issues in fact are a manifestation of larger global issues. Teacher in-service trainers could play a supportive role in designing learning activities that help students and teachers discover that local environmental issues often are a manifestation of global issues.

Organic

An environmental education curriculum should be organic in the sense that there is a good mix of activities and experiences. Teachers should realise that in teaching, content, process and objectives are interdependent. For instance, when the objective is "to improve students' problem solving skills", a video on Greenpeace's actions against whalers will not be adequate. Good environmental educators are able to utilise a range of didactic techniques that could develop a variety of skills essential for environmental problem solving such as: critical thinking, moral reasoning, values clarification, action taking, and holistic thinking. Obviously teacher in-service training institutes need to be able to help teachers in developing these didactic skills.

Learning to learn

Many training programmes have been developed to improve self-regulation knowledge and skills (for a review see Nisbet, 1989; Van Hout-Wolters, 1994). These training programmes differ considerably with respect to the degree of external control being provided by the learning environment. Some programmes are characterised by a relatively high degree of external control while, with other procedures, learners have to rely on themselves to a greater extent (see also Wittrock, 1990). The most fruitful option, however, appears to be a "learning-to-learn" training procedure which is characterised by a gradual transfer of the responsibility for learning processes from the teacher to the learner ("scaffolding") (see Reeve, Palincsar & Brown, 1987; Biemans & Simons, 1992; 1995): students should learn to become more and more responsible for their own learning processes. A gradual withdrawal of external control ("fading"/"scaffolding") is one of the main principles of the so-called "cognitive apprenticeship" approach (see Brown, Collins & Duguid, 1989; De Jong, 1992) which seems to be very useful with respect to participatory learning as well. This also holds true for teacher in-service training situations during which the teacher clearly becomes a learner him or herself.

A learning to learn approach to teacher in-service training and professional development in environmental education, which appears to be highly compatible with the action research approach advocated earlier, could transform teacher training, characterised by a rather high degree of external control, to an in-service programme which allows teachers to shape their own education. When teachers are enabled and equipped to initiate and cultivate their own environmental education, in consultation with school-management, external control can be gradually withdrawn. External facilitation can play a crucial role in the initial phases of environmental education implementation and development. In these early stages, some external support appears to be necessary, while, in later phases of learning, such support can and should be "scaffolded" (see also: Jonassen, 1991). So, "scaffolding" seems to be the most fruitful approach to establish a high degree of self-regulated teacher in-service training, provided that the procedure is based on the actual professional orientation of the teacher. If external support is withdrawn without the teachers being able and prepared to develop their own environmental education, "destructive frictions" and poorer teacher in-service training results are likely to be the result (*cf.* Lohman, 1986; Vermunt, 1992). When the "scaffolding" procedure is tuned to the actual professional orientation of the teacher, however, "constructive frictions" can be created and, thus, the quality of self-regulated learning and teacher in-service training can be enhanced (*cf.* Lohman, 1986; De Jong, 1992; Vermunt, 1992).

CONCLUSIONS

When considering environmental education as both an educational innovation on its own merit and as a means to facilitate educational change, environmental education in-service training cannot ignore a school's policy-making capacity and the professional orientation of teachers. Hence, teacher in-service training in environmental education should not only focus on didactical methods and learning contents, but also on helping teachers broaden their perception of teaching to include curriculum development, developing and utilising teacher in-service training, and contributing to school policy-making.

The current increase within the Dutch education system of school autonomy in developing a professional development programme, does not automatically lead to an increased effectiveness in in-service training and professional development since these areas are not yet a well established practice in schools. Teacher in-service training institutes, more than in the past, will have to assist schools in cultivating strategies for the implementing and developing environmental education which integrate aspects of teacher training, professional development and school policy-making.

Given the importance of involving teachers in the development, implementation and evaluation of educational change or an educational innovation, an action-research approach appears most appropriate for developing environmental education in schools. Teacher in-service trainers should become skilled in facilitating action research in schools. Only through direct interaction with the participants in the school and by 'being there' a teacher in-service trainer is able to tailor a teacher in-service training and professional development programme to the individual needs of teachers and schools.

Effective environmental education in-service training is school-based, purposeful, learner centred, contextual, organic and adopts a learning to learn perspective. Teacher in-service training institutes should become familiar with all these dimensions and assist schools in developing their own needs-based programme for simultaneous teacher in-service training and professional development.

When acknowledging that teachers are also learners themselves, certainly in the context of in-service training, adopting a learning to learn perspective to teacher in-service training can prove to be quite fruitful. Needless to say, we see a relationship between an individual's professional orientation and their learning to learn potential. However, further research is necessary to establish this relationship, particularly in the domain of environmental education teacher in-service training.

REFERENCES

Alblas, A. H., van den Bor, W. & Wals, A. E. J. 1995. The environmental dimension of vocational education. *International Research in Geographical and Environmental Education 4* (3), (in press).

Bergen, Th., Van Amelsvoort, J. & Setz, W. 1994. Het lesgedrag van docenten in relatie tot de vakspecifieke motivatie van leerlingen [Instructional behaviour of teachers in relation to content-specific motivation of students]. *Pedagogische Studiën 71*, pp 256-270.

Biemans, H. J. A. & Simons, P. R. J. 1992. Learning to use a word processor with concurrent computer-assisted instruction. *Learning and Instruction 2*, pp 321-338.

Biemans, H. J. A. & Simons, P. R. J. 1995. *Learning to learn for conceptual change.* Paper presented at the 6th European Conference for Research on Learning and Instruction, Nijmegen, The Netherlands.

Brown, J. S., Collins, A. & Duguid, P. 1989. Situated cognition and the culture of learning. *Educational Researcher 18* (1), pp 32-42.

De Jong, F. P. C. M. 1992. *Zelfstandig leren. Regulatie van het leerproces en leren reguleren: een procesbenadering [Autonomous learning. Regulation of the learning process and learning to regulate: a process approach].* Doctoral dissertation. Tilburg: University of Tilburg.

Elliott, J. 1991. *Action Research for Educational Change: Developing teachers and teaching* Philadelphia: Open University Press.

Hoyle, E. 1980. Professionalization and deprofessionalization in education. In: Hoyle, E. & Megarry, J. (Eds.) *World Yearbook of Education 1980* London: Kogan, pp 42-57.

Jonassen, D. H. 1991. Objectivism versus constructivism: do we need a new philosophical paradigm? *Educational Technology: Research & Development 39* (3), pp 5-14.

Jongmans, C. T. 1994a. Towards a new in-service policy: the case of Dutch agricultural education. *European Journal of Agricultural Education and Extension 1* (2), pp 93-107.

Jongmans, C. T. 1994b. *Van aanbod- naar vraaggerichte scholing van leraren [In-service teacher training: from supply to demand]* Wageningen: Vakgroep Agrarische Onderwijskunde.

Jongmans, C. T. 1995. *Beleidsvoerend vermogen van scholen, professionele oriëntatie van leraren en ontwikkeling van scholingsbeleid [School policy-*

making capacity, teachers' professional orientation and developing in-service policy] Wageningen: Vakgroep Agrarische Onderwijskunde (in press).

Kuhlemeier, H., Van den Bergh, H. & Teunisse, F. 1990. Interne structuur en constructvaliditeit van belevingsschalen voor wiskunde en Engels [Internal structure and construct validity of perception scales for mathematics and English]. *Tijdschrift voor Onderwijsresearch 15*, pp 110-122.

Lewin, K. 1946. Action Research and Minority Problems. *Journal of Social Issues 2*, pp 34-46.

Lohman, D F. 1986. Predicting mathemathantic effects in the teaching of higher-order skills. *Educational Psychologist 21*, pp 191-208.

Nisbet, J. 1989. *Learning to think, thinking to learn: the curriculum redefined* Background paper for the International Conference O.E.C.D., Paris, France.

Procesmanagement Basisvorming 1993. *Richting en ruimte: een elftal voorstellen voor continuiteit in een samenhangend onderwijsbeleid [Room for change: eleven proposals for continuity and coherence in educational policy-making]* Utrecht: Procesmanagement Basisvorming.

Reeve, R A, Palincsar, A S, & Brown, A L. 1987. Everyday and academic thinking: implications for learning and problem solving. *Journal of Curriculum Studies 19* (2), pp 123-133.

Resnick, L. B. 1989. Introduction. In: Resnick, L. B. (Ed.) *Knowing, learning and instruction. Essays in honor of Robert Glaser* Hillsdale: Lawrence Erlbaum Associates, pp 1-24.

Riteco-Slaats, A. 1994. *Leerstijlen in het Middelbaar Beroepsonderwijs [Learning styles in secondary vocational education]* Tilburg: University of Tilburg.

Robottom, I. 1987. Towards Inquiry-based Professional Development in Environmental Education. In: Robottom, I. (Ed.) *Environmental Education: Practice and Possibility* Victoria, Australia: Deakin University Press.

Rosenshine, B. & Stevens, R. 1986. Teaching functions. In: Wittrock, M. C. (Ed.) *Handbook of research on teaching* New York: MacMillan, pp 376-391.

Simons, P. R. J. 1991. *Constructive learning: the role of the learner.* Paper presented at the NATO Advanced Research Workshop on the Design of Constructivist Learning Environments, Leuven, Belgium.

Sleegers, P. J. C. 1991. *School en beleidsvoering. Een onderzoek naar de relatie tussen het beleidsvoerend vermogen van scholen en het benutten van de beleidsruimte door scholen [School and policy-making. A study of the relation between school policy-making capacity and real policy-making]* Doctoral dissertation. Nijmegen: University of Nijmegen.

Stapp, W. B. & Wals, A. E. J. 1994. An action research approach to environmental problem solving. In: Bardwell, L. V., Monroe, M. C. & Tudor, M. T. (Eds.) *Environmental Problem Solving: theory, practice and possibilities in environmental education* Troy, Ohio: NAAEE.

Van Amelsvoort, J., Bergen, Th., Lamberigts, R. & Setz, W. 1993. *De invloed van de kwaliteit van de instructie op de motivationele oriëntatie en de schoolcarrière van leerlingen [The influence of instructional quality on students' motivational orientation and educational career]* Nijmegen: KUN/ITS.

Van Hout-Wolters, B. H. A. M. 1994. "Leren leren" in het onderwijs: ontwikkelingen en problemen ["Learning to learn" in education: developments and problems] *VELON Tijdschrift 15* (3), pp 3-7.

Vermunt, J. D. H. M. 1992. *Leerstijlen en sturen van leerprocessen in het hoger onderwijs: naar procesgerichte instructie in zelfstandig denken [Learning styles and regulation of learning in higher education: towards process-oriented instruction in autonomous thinking]* Doctoral dissertation. Lisse: Swets & Zeitlinger.

Wals, A. E. J., Beringer, A. & Stapp, W. B. 1990. Education in Action a Community Problem Solving Program for Schools. *Journal for Environmental Education 21* (4), pp 13-20.

Wals, A. E. J. 1993. Critical Phenomenology and Environmental Education Research. In: Mrazek, R. (Ed.) *Alternative Paradigms in Environmental Education Research* NAAEE Monograph Series Troy, Ohio: NAAEE.

Wals, A. E. J. 1994a. Action Research and Community Problem Solving: environmental education in an inner-city. *Educational Action Research 2* (2), pp 163-183.

Wals, A. E. J. 1994b. Pollution Stinks! *Young adolescents' perceptions of nature and environmental issues with implications for education in urban settings* De Lier: Academic Book Center, ABC.

Wals, A. E. J. 1994c. Action taking and environmental problem solving in environmental education. In: Jensen, B. B. & Schnack, K. (Eds.) *Action and action competence as key concepts in critical pedagogy* Didaktische studier, Volume 12 Kopenhagen: Royal Danish School for Educational Studies.

Wittrock, M. C. 1990. Generative processes of comprehension *Educational Psychologist 24* (4), pp 345-376.

CHAPTER 3

LEARNING TO EDUCATE: ENVIRONMENTAL EDUCATION IN TEACHER EDUCATION IN SCOTLAND

Tony Shallcross

> *"Adding things to the curriculum is all very well but it is not enough. You've got to change the way teachers talk to you about it."*
>
> Fifth-year Scottish school student talking about environmental education (Scottish Office Environment Department, 1993)

PROLOGUE

Scotland has a proud academic history in environmentalism, but not a strong applied tradition in environmental education, especially in initial teacher education (ITE). Recently, however, two Scottish reports on environmental education have attracted international interest and acclaim. 'Environmental Education' written by Her Majesty's Inspectors of Schools in Scotland (HMIS) in 1974 was a pioneering document. Despite the positive response that this report received, the proposals contained within were never implemented. Contemporary local government and curriculum reform may explain its lack of implementation.

History may be repeating itself in the 1990s. *Learning for Life*, Scotland's National Strategy for Environmental Education (1993), has also attracted very favorable international and domestic comment. It has been described as comprehensive, thorough, thoughtful and practical. Yet it has taken more than two years for 'A Scottish Strategy for Environmental Education' (The Scottish Office, 1995), the official statement of intent from the Secretary of State for Scotland to be published, even though *Learning for Life* was commissioned as the Scottish Office response to *Agenda 21.* Is *Learning for Life* like 'Environmental Education' fated, or perhaps predestined to become another erudite, heroic failure?

The report is unlikely to fail. It has at the very least, been a catalyst which has set changes in motion in teacher education which make it highly unlikely that the *Learning for Life* initiative will fail in this sector, despite what appears as a lukewarm official response from the Secretary of State for Scotland. Despite, or perhaps because of, the tardy response from the Scottish Office, *Learning for Life* may have been the stimulus for a number of initiatives in ITE, not the least being the strong contacts that it has fostered between the teacher training institutions in Scotland. The existence of a national strategy has fundamentally changed the climate of environmental education in Scotland. It could be said to have induced a condition of curricular warming!

Two main questions have to be answered in relation to environmental education in ITE in Scotland over the next few years. The first is whether education for sustainability, which must figure prominently in any post modern prescription for environmental education, requires a different type of teacher? The second question is which model of change do institutions of ITE adopt if education for sustainable living is to feature prominently in formal education in Scotland? Most evidence indicates that a bottom up model is likely to be the only viable approach.

Scotland has its own educational system. There is no statutory national curriculum, but there are a series of guidelines for the 5-14 curriculum and a single national examining system administered by the Scottish Examinations Board (SEB) for the upper secondary years (see Table 3.1).

Teacher education in Scotland also differs from the rest of the UK in a number of ways, some of which will be examined later. But one very significant difference is the existence of the General Teaching Council (GTC). All courses of teacher education have to be approved by the GTC. But there are similarities with the rest of the UK. Scottish teacher education has been overtaken by the competence revolution, but neither the Scottish Office Education Department (SOED) or the GTC have recognised competence in environmental education is a necessary skill for the teachers of the next century.

Table 3.1 National curriculum structure in Scotland 1993

Age range	*Curriculum*
5-14	Environmental Studies Expressive Arts
P1-P7	Language Mathematics
S1-S2	Religious and Moral Education
15-16	Standard Grade Discrete Subjects: Geography, History, English *etc.*
S3-S4	Scottish Examinations Board (SEB) Syllabuses
17-18	Higher Grade and 6th Year Studies (Academic)
S5-S6	SEB Short Courses (40 hours) Scotvec Vocational and Short Academic Courses

CONCEPTIONS AND CONCOCTIONS; WORKING DEFINITIONS OF ENVIRONMENTAL EDUCATION IN SCOTLAND

One very positive aspect of *Learning for Life* has been that it has helped to produce some degree of consensus about the definition of the environment, a clear prerequisite if a coherent view of environmental education is to emerge in Scotland.

> *"The environment refers to the whole environment in which people live and work, natural and built, local and global, social and cultural."* (Smyth, 1995).

Smyth recognises that this definition of the environment was not widely appreciated outside environmental education circles before Learning for Life was published. However current environmental education practice is less consistent and coherent than his definition might imply. Even within the broad community of teacher educators environmental education is still confused with environmental studies (a similar confusion appears at a number of points in the Secretary of State's statement of intent).

Environmental education in ITE consists mainly of ecology, fieldwork and learning about environmental problems. Like the rest of formal education there is a largely cognitive and cerebral focus in environmental education in many aspects of ITE courses in Scotland. There are many in ITE who would see environmental education as nature conservation, concerned with learning in and about the environment, which means that even fieldwork has an instrumental function in which the environment becomes an object of study used to enhance students' understanding. There are others who would argue for the integration of conservation and development education as education for sustainability, sustainable development or sustainable living; definitions consistent with *Learning for Life*'s concerns for sustainability. What is generally neglected in these approaches is education for the environment, learning to care, through fieldwork and other approaches which have an affective as well as an interventionist role focused on environmentally sustainable action. The transformational nature of approaches to environmental education, such as socially critical education or deep ecology, is also rarely addressed in ITE courses in Scotland.

However a more radical definition of environmental education does appear to be emerging from those in Scotland's community of teacher educators who are directly concerned with the delivery of courses in environmental education. This definition emphasises personal responsibility within a whole institution approach to environmental education, in which environmental education is a permeating cross-curricular theme which is the responsibility of all subjects. Understanding the local and global impact of decisions and acting on these understandings are also seen as essential ingredients of this definition. Clearly all these strands have clear connotations and implications for the teaching of values and attitudes. For this group of teacher educators anything less than this conception would represent a shallow definition of environmental education, more consistent with environmental studies, rather than the more holistic approach which should typify environmental education.

This more radical view of environmental education may be emerging for pragmatic as well as philosophical reasons. Most teachers are not committed environmental activists. Many of them would see biospheric egalitarian ethics as a green fundamentalist threat to the anthropocentric ethics that dominate much western thinking; a metaethical stance which has a strong foundation in Christian and biological conceptions of humanity's rights with regard to the use of global resources. Yet these same teachers would all wish to be associated with good practice.

The devolving of budgets to schools raises another pragmatic problem for environmental education in in-service training (INSET), that will be examined later but it also brings with it more frequent inspection of schools. There are many anecdotal benefits of successful environmental education in Scottish schools - Scottish Consultative Council on the Curriculum, SCCC (1993), such as reduced energy costs, higher self esteem in pupils, more pleasant working surroundings which all make environmental education a very attractive innovation for schools. Many of these indicators are mentioned in recent Scottish Office publications (SOED, 1991, 1992a, 1992b).

Unfortunately not only are the SOED not going to require that environmental education is a specific competence in ITE neither will it be a compulsory area for school inspections. The following reference from the Secretary of State's statement of intent appears to put inspection in a reactive rather than a proactive role in relation to environmental education. The extract below also shows how environmental studies and environmental education can be conflated.

> *"Like other aspects of schools, their activities in environmental education fall to inspection. Environmental education is incorporated in the environmental studies guidelines as a key part of the strand 'developing informed attitudes'. As schools implement the 5-14 guidelines, so too will HM Inspectors of Schools find increasing evidence of attention being given to environmental education and that will lead to appropriate references in published reports."* Scottish Office Environment Department (1995).

The pragmatic message that should be coming through ITE and INSET is that environmental education makes for better schools. Many may see this as a concocted conception but it is one which the majority of teachers would find attractive and non threatening. When assessment is made in the future about education for sustainability schools will be judged by the quality of their environmental actions not by the theoretical purity of their mission statements. Unfortunately the Scottish Office statement of intent does not appear to require that schools have even have a policy for environmental education, let alone that as we have already seen that environmental education is reflected in their practice.

However, many anecdotal examples of good school practice emerged from the OECD Environment and National Schools Initiative (ENSI) in which 15 Scottish schools participated. These case studies are becoming increasingly used in Scottish ITE; perhaps their use will lead to the acceptance of a more coherent definition of environmental education throughout Scottish formal education. But there is little hard research evidence to highlight the benefits of adopting whole school approaches; it is an area of research that the SOED have yet to prioritise. What better way is there of demonstrating the efficacy of whole institution approaches than to develop these in institutions of teacher education? While a strong feature of environmental education courses in both ITE or INSET should be the advertising and analysing of good practice, engaging in good practice is a much better form of advocacy.

ENVIRONMENTAL EDUCATION IN TEACHER EDUCATION IN SCOTLAND

There are six institutions involved in ITE in Scotland, the largest is the Faculty of Education of the University of Strathclyde in Glasgow, followed by Moray House, the Institute of Education of Heriot Watt University, Edinburgh, Northern College of Education based in both Aberdeen and Dundee, St. Andrew's College of Education in Glasgow which is a Catholic institution, the Faculty of Education of the University of Paisley at Craigie and the University of Stirling. INSET provision is much more complex. All these six institutions are major INSET contributors in environmental education, but other organisations such as the regional education authorities, NGOs such as the World Wide Fund for Nature (WWF), Royal Society of the Protection of Birds (RSPB), the Scottish Development Education Centre, many ranger services, field centres and some schools and further education colleges are also providers of INSET in environmental education.

Although there are obvious areas of neglect in the current provision for environmental education in ITE in Scotland which have already been highlighted, formal education still has a major role to play in the quest for sustainability. One of the main reasons for environmental education's limited success in Scottish schools, has been the failure to recognise the crucial role of teacher education in the process of developing sound environmental action, a point highlighted in many of the responses made to *Learning for Life* (Smyth, 1995a). Official statutory support for environmental education in Scotland would have been a great fillip, but progress can still be made despite the absence of the high degree of support for environmental education in schools that exists, for example, in the Netherlands. If the majority of probationary teachers entering Scottish schools are proficient and enthusiastic environmental educators, most school pupils are likely to be exposed to good environmental education practice at some stage in their schooling. While this provision alone will not guarantee sustainable behaviour in Scotland it is a necessary condition for the achievement of sustainable development.

At the same time as this pre-service innovation, senior managers in schools should be the targets for INSET courses in environmental education. Unfortunately, devolving budgets to schools without a statutory requirement or strong political support for environmental education makes it unlikely that environmental education will be seen as a priority for INSET funding by most headteachers. The declining power of local and regional educational authorities in the face of local government reform in 1996 has also emasculated the role of advisors in providing or commissioning INSET courses in environmental education at a regional level.

In 1993, Lord James Douglas Hamilton, Scottish Office Minister of Education observed that the publication of curriculum guidelines had removed the need for the advisory service to provide curriculum support. As advisors turn their hands to quality assurance, the new education authorities in Scotland will be less likely to give an INSET lead in areas such as environmental education. The day to day realities of the SOED's 5-14 curriculum guidelines (see Table 3.1) are much more likely to attract INSET funding from primary schools than cross-curricular themes such as environmental education. Without official encouragement a strong environmental ethos is unlikely to develop quickly or with any degree of uniformity in Scottish education. But the Dutch experience does show that devolution of budgets and strong central support for environmental education are not incompatible (Jansen et al, 1995). In the absence of an official requirement and strong central support for environmental education in Scotland a bottom up approach involving ITE is a very attractive and realistic alternative. Scottish institutions of teacher education have responded to this INSET dilemma in many ways. One is to include a module on environmental education in the Post Graduate Certificate in Environmental Studies, a distance learning course produced by a consortium of all the institutions of teacher education in Scotland, with the exception of the University of Stirling. The whole course has recently been rewritten in the light of the SOED's 5-14 Environmental Studies (1993) and *Learning for Life* documents. In rewriting the course, care has been taken to address as wide an audience as possible including primary and secondary school teachers, senior managers in schools, rangers, education officers with NGOs and staff working in field centres. Another response has been the development of environmental education modules in modular masters degrees in at least three institutions, two of which are also seeking to develop masters programmes in environmental education. Further, there is a well established post graduate environmental education course offered by the University of Strathclyde but this is aimed almost exclusively at the overseas market (Dunlop, 1995).

In a system in which INSET cannot guarantee coherence, quality assurance or progression in its environmental education provision, the role of ITE is crucial. New teachers should be enthusiastic, confident practitioners of environmental education when they graduate; for in the current educational climate we cannot assume that as teachers they will be exposed to these skills in their professional careers.

Environmental education in ITE or INSET levels usually concentrates on one, occasionally two, of the four school based roles, shown in Figure 3.1. Much teacher education overlooks the first role; the teacher as individual and role model. Research is going on in ITE in Scotland into the environmental values of students and how these affect environmental behaviour (Shallcross and Wilkinson, 1994). What this research has shown so far is that the majority of students sympathise with environmentally sustainable action but for a variety of reasons do not always act consistently with these beliefs; what Aristotle referred to as incontinent behaviour.

Figure 3.1 Some differing roles of teachers *in* school

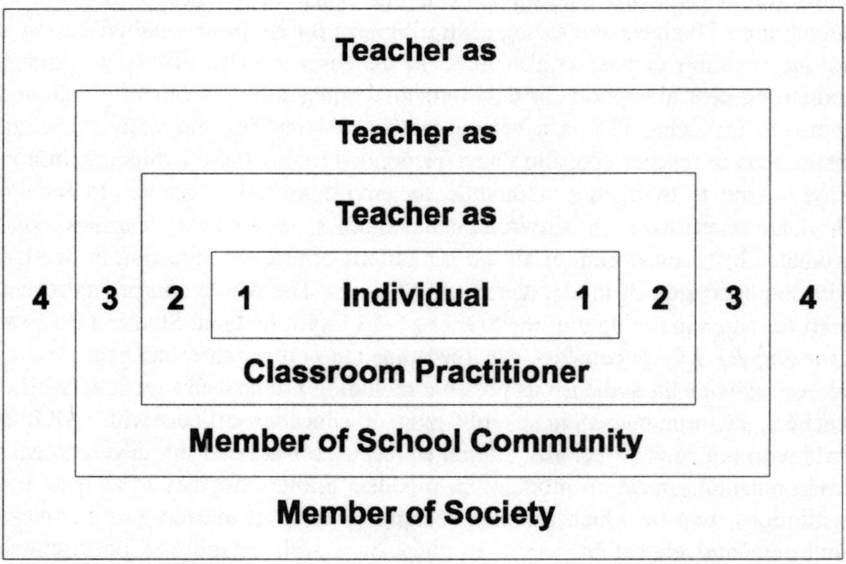

Some of these students are required to practise environmental education on placement. These students will also be exposed to a more integrated cross-curricular provision for environmental education in an institution which is also developing its own environmental strategy. They are also introduced to the teaching of controversial issues through the techniques of values education. One objective of the research is to see what influence an environmental education course which focuses on pedagogical skills within a permeating cross-curricular approach, supported by a developing whole institution approach, has on incontinence in comparison with the more traditional content based courses outlined earlier.

The arrangements which are made for environmental education in ITE differ from one institution to the next, as one would expect, but they are also very dependent on the stage of schooling for which students are training. Primary teachers can qualify through two routes, the four year Bachelor of Education degree (B.Ed. Primary) or the one year Post Graduate Certificate of Education (PGCE Primary) the latter course would follow a first degree. The majority of students follow the four year course which obviously gives more scope for incorporating environmental education than the one year PGCE course. On the PGCE courses students will spend at least half their time on placement in schools, approximately a quarter of the four year B.Ed. course is spent in school.

The majority of students training as secondary teachers follow a one year PGCE course. There are some students who follow four year B.Ed. courses in subjects such as physical education and technology. Students at the University of Stirling can achieve qualified secondary teacher status by following a course in education concurrently with their main degree course.

The PGCE Secondary course has been selected by the SOED for mentoring. While mentoring will not start until at least 1995-96 there is no doubt that the SOED eventually would like to see 22 weeks of this course based in schools with the remaining 14 weeks in institutes, faculties or colleges of education. The consequences for environmental education are stark, the amount of time that environmental education can attract out of a maximum contact time of 110-140 hours is minimal; in some institutions of ITE non-existent. This is a serious situation given that most interested commentators agree that the provision for environmental education is weaker in Scottish secondary than primary schools.

There are a number of models used to teach environmental education on B.Ed. courses in Scotland; these range from full modules, to half modules or parts of modules to selected permeating inputs. The emphasis on the acquisition of pedagogical skills is an increasingly common design feature of new or revised environmental education courses in B.Ed. degrees in Scotland. These are not courses designed to raise environmental awareness, they are courses on good practice, based on the premise that the most effective environmental education in primary schools is one based on a whole school approach. Environmental content is covered in compulsory environmental studies courses, in electives and in some cases as a permeating theme.

WHAT SORT OF SKILLS AND QUALITIES DO TEACHERS NEED TO DELIVER EDUCATION FOR SUSTAINABLE LIVING?

We have to recognise that informing and or changing attitudes to effect sustainable lifestyles is fundamentally an exercise in values education. At the heart of this process, as Figure 3.1 shows, are the values of individual teachers. The first, perhaps, the most important task for teachers is to be both a professional and individual role model; a task which is arguably more important with younger pupils. Students and teachers are role models whether they like it or not, but perhaps the significant influence that educators offer at the moment is the subconscious socialisation of hypocrisy. If environmental education continues as an essentially cognitive and cerebral activity, in which educational institutions fail to practise what they teach, they may be giving pupils a clear subliminal message through the informal curriculum that it is socially acceptable to advocate one thing and do another.

> *"Dewey would have found it interesting that an institution may be keen to put on courses on ecology, energy and the like while ignoring its own environmental impacts. For example, course handouts may not be printed on recycled paper, and classes may be held in poorly insulated energy wasting buildings."* (Brennan, 1991).

It is not surprising therefore if adults justify their own lack of environmental action by highlighting the hypocrisy apparent in wider society.

> *"I think you would do more if you thought you were achieving something, but as an individual you don't really feel that you are doing anything. Then you hear about some mountain of something in Europe and you wonder 'what is the point?' You try and conserve energy at home, and cut down and then you go into town and see all these wasted lights in shops. You think 'why bother?'"* (Corrado & Nove, MORI, 1995).

Brennan illustrates the importance of whole school approaches to environmental education. If teachers are going to develop these approaches they need skills which many of them do not currently possess. Institutional auditing is one such skill. Monitoring practices such as energy consumption features in environmental education courses in at least two institutions of ITE. Students acquire the auditing skills which are the first stage in planning for institutional change. Exposing students to examples of good practice as well provides them with a range of models for the planning of whole school approaches. Being aware of theories and strategies of change is also important in enabling students to contribute to, and perhaps initiate, whole school approaches when they become practising teachers.

The focus on active and cognitive educational strategies in most schools appears to be based on the assumption that positive feelings for the environment will follow by osmosis from learning in and about the environment. But if attitudes are to be informed, developed or changed much greater emphasis has to be given to learning for the environment, which in turn requires the inclusion of specific strategies for affective teaching and learning in the environmental education curriculum. Attitudes have active, affective and cognitive, components. Environmental education takes place in, for and about the environment. The congruence could not be closer. But most schools and institutions of ITE do not give the affective domain sufficient attention in their environmental education programmes. We cannot require teachers to develop informed attitudes without exposing them to all the means of achieving these.

The obvious question is how we are to redress the deficiency in affective education provision in ITE? Didactic approaches which focus on global environmental problems appear to be moribund and possibly counter productive; a more appropriate strategy is to develop skills based courses with an implicitly ecological philosophy. Teachers who experience affective teaching strategies are much more likely to use these inside and outside the classroom. There should obviously be scope for students to learn more about the environment as part of ITE, but the environmental education component of ITE courses should focus on teaching, research and evaluation skills. As the guidelines for teacher training put greater emphasis on the generic, so the delivery of skills will become one process on which ITE is predicated.

The approaches to affective education embodied in values education and sensitisation are particularly relevant to teacher education. Sensitisation by heightening sensory perception of the environment affords the best opportunity for children to achieve a sense of joy and wonder about the environment; these techniques are taught on most Scottish ITE, environmental education courses.

Raising the status of emotional discourse by exposing all trainee teachers to affective education programmes is one way in which environmental education can permeate the whole curriculum. In practical terms this can also be addressed by making reference to feelings legitimate in all branches of the curriculum through the wider use of story, metaphor, poetry, drama and other techniques.

Education for the environment should concentrate on creating a sense of awe and wonder about nature. Most students feel confident in teaching in and about the environment, but few are as confident about teaching for the environment. Without addressing approaches which focus on the affective component of attitudes how can we be confident that the next generation will care for the environment and express this care through sustainable action? The environmental insights gained from drama and poetry are every bit as valuable as purely ecological understandings and many affective techniques can be incorporated into mainstream curricular practice in order to achieve these affective outcomes. Education in the environment should also be concerned with good environmental practice in the school and its grounds as well as fieldwork outside the school.

Children will cherish what they value, but a strong emotional attachment to the environment must be built on a sound understanding of ecological processes. In this way children will feel that their ownership of the environment is intellectual as well as material and emotional. Responsible action will be enhanced by environmental understanding, especially if children feel that their actions can affect the quest for sustainability. Introducing ITE students to curriculum planning tools such as the one in Figure 3.2 below is one way of incorporating these concerns and also dealing with the thorny issue of sequencing (based on Lyle, 1994)

Pedagogical skills cannot be taught in a vacuum. Using teaching materials which illustrate approaches to values education and other skills in case studies drawn from Scotland demonstrates the use of local examples in environmental education. But it goes further than this, it actually addresses another area of neglect. Many school pupils in Scotland are not aware that Scotland has less than 2% of its Caledonian pine forest left. They do not recognise that many parts of the Scottish Highlands are severely environmentally degraded, perhaps unsustainable landscapes. Yet many of these same pupils are knowledgeable about tropical rainforest. Starting with the local is a traditional application of the spiral curriculum studying the immediate as a way of transferring and applying concepts to more distant locations, thereby facilitating sound comparisons.

This is not to say that pupils and students should not be made aware of environmental issues in distant places, but there are dangers if pupils are not

Figure 3.2 Environmental education curriculum planning tool

Stage 1 **Ecological understanding**	How do ecosystems function? Gaia hypothesis. People as part of nature, not apart from and superior to it. The relationship between social, economic and environmental systems.
Stage 2 **Aesthetic appreciation**	The beauty of the natural, social and built worlds. The importance of affective learning. Educating the emotions and feelings. Learning to care for nature, other people and distant populations.
Stage 3 **What is happening to natural and social systems?**	The planet's natural and social systems. Is there an ecological crisis? Is world poverty endemic and the result of process of unequal exchange which favours the advanced countries?
Stage 4 **What are the consequences?**	Now and in the Future.
Stage 5 **Why is it happening?**	Which current practices, policies, processes and beliefs are contributing to the issue or alleviating it?
Stage 6 **What can WE do about it?**	In school, within the community, nationally, through the democratic process, globally through charitable actions and/or ethical consumption.
Stage 7 **What are WE doing?**	Audit of possible actions followed by an action plan as part of a whole school approach.

aware of some of the issues which arise in their own society. The more distant the example the more distant the responsibility. Studying issues such as tropical rainforest can emasculate. Pupils understand the dangers but either because of the scale of the problem or their distance from the issue feel powerless to do anything about it (Froud, 1994). The more local the example the more likely it is that pupils and adults can do something and thereby gain new insights and powers.

> *"You tend to be interested in what affects your local area. And what is going to happen to your children."* (Corrado & Nove, MORI, 1995).

Clearly, many environmental issues are complex. They also involve conflicts between different values and ideologies. At the same time curricular guidelines such as the SOED 5-14 Guidelines charge teachers with the responsibility for developing informed attitudes towards the environment. Learning for Life requires changed action as the extract below indicates. Therefore environmental educators are concerned with values in two senses. Firstly informing and developing the values of individuals. Secondly giving pupils empathy and insight into the values of others. Values clarification skills enable pupils and students to scrutinize their own values. Values inference skills examine the values of other individuals and groups. Values analysis skills which can be used to evaluate their own values and those of other individuals and groups. Here is a classroom application that ITE students need to master.

> *"Our environment will not stand up to the pressures that are now being put on it by increasing human population and activity. We recognize that the most effective way to ensure its protection and enhancement in the long term is to bring about change in people's attitudes and behaviour towards it."* (Scottish Office, 1993).

However, the principles of values education do create a predicament for schoolteachers. On the one hand official documents such as *Learning for Life* and the '5-14 Environmental Studies Guidelines' advocate changing and informing attitudes. On the other hand inculcating one's own ideology in the classroom is not deemed to be politically correct. When teaching about controversial issues which have any political dimension to them the operational curriculum should be value fair. Under these circumstances how can schools ensure that the attitudes that children develop are those which will lead to positive environmental action? It is through the informal curriculum that schools give their strongest moral and ethical lead in the actions that the school approves and the ethos it establishes. Hence the centrality of the whole school approach to ITE in environmental education (Northern Ireland Curriculum Council, 1994).

Yet it is only by recognising that our current models of environmental education exist in a world in which development dominates our thinking that we can identify the nature and scale of the educational changes required to secure a sustainable future. Perhaps the major cognitive breakthrough is for those of us who live in the North to recognise that the major environmental threat to the planet could very well be advancing rates of consumption in the North and its consequent pollution, rather than rates of population increase in the South. After all one of the major arts of education is to make the ordinary seem extraordinary! Students in ITE have to be able to evaluate more radical and transformational models of environmental education in their training and given the opportunity to incorporate these into their practice.

HOLISTIC APPROACHES IN IVORY TOWERS?

To modify an adage, institutions of ITE have to practice what they teach. HE institutions and their lecturers have to become role models of good practice. We cannot advocate whole school, cross-curricular, thematic approaches and the importance of personal responsibility and then eschew these in ITE practice. As Smyth (1995b) points out it is unrealistic to teach students through a disciplinary mode and expect them to forge inter-disciplinary and holistic links for themselves, this statement is even more pertinent when we consider the primary schools and classrooms in which many of these students will teach. Various schools of environmental education such as socially critical educators and deep ecologists advocate holism; the Gaia hypothesis is perhaps the most articulated epistemological version of this message. Yet how frequently do we encounter integrated, holistic approaches to environmental education especially within institutions of higher education?

There are some signs that this almost self evident truism is beginning to permeate the collective consciousness of Scotland's community of environmental educators in ITE. A group has been established to examine the extent to which all the institutions of teacher education in Scotland can collaborate in the provision of the environmental education in ITE while at the same time the majority of these institutions are embarking on the development of environmental strategies.

Institutions of teacher education must set the lead in developing high quality whole institution approaches to environmental education which permeate all aspects of their own practice; this is by far the fastest way of aiding good environmental education in schools. Whole institution policies are the best, arguably the only genuine form of cross-curricular permeation of environmental education in any educational institution. At least three institutions involved in ITE have initiated an environmental policy. At least two are developing their own centres for environmental education. Most colleges are involved in developing their grounds for environmental education purposes, an initiative which has recently attracted funding from Scottish Natural Heritage (SNH). It is through experiencing this whole institution approach that the teachers of the twenty first century should learn how to educate for sustainability.

REFERENCES

Brennan, A. 1991. Environmental Awareness and Liberal Education. *British Journal of Educational Studies 39* (3), pp 279-295.

Corrado, M. & Nove, A. MORI, 1995. *Environmental Education. Is The Message Getting Through?* Edinburgh: Scottish Office Central Research Unit, HMSO.

Dunlop, J. 1995. Promoting Training in Environmental Education the Environmental Education Certificate Course on the Jordanhill Campus, University of Strathclyde, Glasgow. In: Leal Filho, W., MacDermott, F. & Murphy, Z. (Eds.) *Practices in Environmental Education in Europe* Bradford: European Research and Training Centre on Environmental Education.

Froud, K. 1994. *Environmental Education: Teaching Approaches and Pupil Attitudes* Slough: National Foundation for Educational Research (NFER).

Jansen, B., Reehorst, E., & Delhaas, R. 1995. Action Research Networking and Environmental Education in the Netherlands: a challenging and successful action research strategy for educational reform. *Educational Action Research 3* (2), pp 195-209.

Lyle S. 1994. *Environmental education - the implications of the Welsh advisory paper,* Unpublished paper given to the Conference of the Association of Teacher Educators in Europe, Prague.

Northern Ireland Curriculum Council 1994. *Environmental Education A Guide For Teachers* Belfast: Northern Ireland Curriculum Council.

OECD Environment National Schools Initiative 1994. *Environment and School Initiatives (ENSI) Project in Scotland The National Report* Dundee: Scottish Consultative Council on the Curriculum (SCCC).

Scottish Education Department 1974. *Environmental Education; a Report by HM Inspectors of Schools (*The Gilbert Report) Edinburgh: HMSO.

Scottish Office Education Department 1991. *The Role of School Development Planning in Managing School Effectiveness* Edinburgh: HMSO.

Scottish Office Education Department 1992a. *Using Performance Indicators in Schools Self-Evaluation* Edinburgh: HMSO.

Scottish Office Education Department 1992b. *Using Ethos Indicators in School Self-Evaluation* Edinburgh: HMSO.

Scottish Office Education Department 1993. *Environmental Studies: 5-14 Guidelines* Edinburgh: HMSO.

Scottish Office Environment Department 1993. *Learning for Life* A *National Strategy for Environmental Education in Scotland* Edinburgh: HMSO.

Scottish Office Environment Department 1995. A *Scottish Strategy for Environmental Education* Edinburgh: HMSO.

Shallcross, A. & Wilkinson, G. 1994. Sustainable Values the Basis of Teacher Education for Sustainability. In: *Proceedings of the Conference of the Association of Teacher Educators in Europe, Prague*, in print.

Smyth J. 1995a. A *National Environmental Education Strategy for Scotland.* In: Leal Filho, W., MacDermott, F. & Murphy, Z. (Eds.) *Practices in Environmental Education in Europe* Bradford: European Research and Training Centre on Environmental Education.

Smyth J. 1995b. Environment and education: a view of a changing scene. *Journal of Environmental Education Research 1* (1), pp 3-20.

CHAPTER 4

ENVIRONMENTAL EDUCATION IN TEACHER TRAINING IN PORTUGAL

Lídia Máximo-Esteves

INTRODUCTION

Collating the multiple initiatives regarding environmental education, that have taken place in Portugal is not an easy task. Neither is it easy to describe the situation regarding teacher training in environmental education. There are several reasons to explain such difficulty:

(1) The peculiar introduction of environmental education into the Portuguese educational system.

(2) The lack of a well defined, nation-wide policy concerning environmental education in teacher training, which leads to a lack of systematic planning and to spontaneous and occasional training.

(3) The existence of several institutions able to give teacher training, as well as the autonomy to elaborate teaching programmes, in pre-service and in-service teacher training[1].

Given this situation, it is extremely difficult to estimate the number of environmental education teacher training courses taking place in Portugal. Further, it is as equally difficult to provide an accurate description of these courses as it is to quantify them. However, this paper seeks to present an overview of teacher training for environmental education in Portugal.

---

1 This decentralisation allows for better solutions locally, but makes it more difficult to estimate the range of activities involved.

THE ORIGINS OF ENVIRONMENTAL EDUCATION IN PORTUGAL

It was in 1969 that, by request of the United Nations, the first Commission to discuss environmental issues at an international level was created in Portugal. As a result, one of the first official documents about international environmental issues was developed in preparation for the Stockholm Conference held two years later in 1972.

In solidarity with the close attention paid to the environment throughout Europe, Portugal has ever since participated in international meetings and conferences. As part of this trend, Portugal has created suitable governmental structures and a legislative and institutional framework for environment protection. From early on in this process, plans were made to provide widespread education about the issues involved. The expression "education of the population" was used broadly and included non-formal as well as formal education.

The Comissão Nacional do Ambiente (CNA) (National Commission for Environment) was created in 1971 with the aim of 'intensifying and co-ordinating in-country activities' related to environment. This commission was formed by representatives of several government Ministries and also by a number of individual practitioners. The CNA turned out to be a central organisation for the co-ordination and promotion of initiatives connected with the environment and environmental education, both at national and international levels. It took on the role of informing and educating the Portuguese population as regards environmental issues, and also acted as advisor to the Ministry of Education.

Since its foundation and until 1974 (the year of the April Revolution), the CNA worked towards the following general aims:

(1) to promote the exchange and dissemination of environmental information, especially the outcomes of International Conventions;

(2) to promote the introduction and use of environmental issues in teaching programmes (Evangelista, 1992).

During this period the CNA was much more prominent as a provider of information than of training. At this stage the environment was regarded as "a novelty or as means of social and political criticism and not as an educational tool" (Evangelista, 1992).

The April Revolution in 1974[2] had immediate effects on Portuguese education; the basic principles of the educational system in the previous regime were abolished, the curricula were suspended and alternative arrangements were made while new programmes were being prepared. As a result the CNA experienced a period of uncertainty regarding an exact role.

Despite this, however, the CNA had some impact on the alterations made in the environmental education content of curricula for teacher training at all levels. This was achieved through collaboration with the Direcções Gerais de Ensino do Ministério da Educação (Teaching Departments of the Education Ministry). New disciplines were created - they presented updated information on environment issues; some discipline programmes were reorganised and environmental issues were introduced in order to establish an interdisciplinary approach. To this end, tasks undertaken by CNA included:

- the creation of a documentation centre on environmental education for teachers and students;
- the refinement of pedagogical-didactic materials for environmental education, in order to give support to teachers and students who requested them;
- the organisation of meetings with students of all levels, according to the request of schools to promote environmental awareness. It had organised talks, films and slides were exhibited, and books were made available. Many of these books, like "The World is Our Home" was widely disseminated among the student population;
- the preparation and delivery of teacher training courses for all teaching levels, with the aim of providing knowledge about the new themes involved and teaching approaches which environmental education required.

In 1978, the CNA organised a number of courses for supervisors in charge of directing primary school teachers - Meio Fisico e Social (Physical and Social Environment). From 1976 to 1983, two hundred courses of three days each took place, with the participation of over 10 000 teachers from all teaching levels (IPAMB Report, 1993).

After a period of constant restructuring, the CNA was dissolved in 1983. Its tasks were temporarily transferred to one of the other Ministries, and in the following year these tasks were further transferred to the National Service of Parks and Nature Resources.

Then in 1987, another organisation, with similar functions in relation to training, information and co-ordination of environmental education, was created. Known as the Instituto Nacional do Ambiente - INAMB (National Environment Institute), it has since been replaced by the Instituto de

2 The 25th of April 1974 saw the end of a political dictatorship and the beginning of a democratic regime. This lead to a period of political, economical and administrative instability in Portuguese society which was reflected in the organic structures. It was some time until the State and Society reorganised themselves.

Promoção Ambiental - IPAMB (Environment Promotion Institute) which came into force in 1993.

The INAMB was established with the 1987 Environment Law. The functions of this organisation were to inform and educate citizens through the "study and promotion of special projects on environmental education". In order to achieve this goal, INAMB undertook the following actions:

- Support was given to several environmental education projects developed in schools, in conjunction with the Direcções Gerais de Ensino do Ministério da Educação (Teaching Departments of the Education Ministry).
- The goals and methodology that were tested and evaluated through these projects were disseminated through a free leaflet sent to schools.

The main aim behind both actions was to raise awareness among teachers of the environment and environmental education.

ENVIRONMENTAL EDUCATION IN THE BASIC LAW OF THE EDUCATION SYSTEM

The history of the introduction of environmental education to the Portuguese education system illustrates that this task was achieved gradually through a succession of events. Initially it depended more upon the initiative of informed teachers than to courses prepared by organisations for this purpose

In order to better understand this process, it is necessary to have an overview of what was happening with the introduction of environmental education at all teaching levels, after 1974[3].

(1) At kindergarten level there was no formal curriculum. To-date this situation remains unchanged. Pedagogical activities for environmental education generally occur sporadically according to the level of awareness of the teachers involved.

(2) A new subject, Meio Fisico e Social (Physical and Social Environment), was introduced into the 1st cycle of Basic School. The main goal was to make pupils aware of the environment and environmental issues, particularly concerning respect for and protection of the environment.

(3) New environmental themes were developed in the 2nd and 3rd cycles of Basic School and also into Secondary level education. Not all, however, were successful, and some were even removed, partially due

to the lack of well trained teachers, in spite of the efforts made by the institutions in charge of training.

The first government documents on environmental education in the curriculum were published at the beginning of the second decade following the April Revolution. They created the legal structure needed to enable the necessary alterations in the educational system to take place as well as providing solid guidelines for the selection, training and career progression of teachers.

In order to understand the importance of this initiative it is necessary to comment upon the main principles which lay behind the new laws that still govern the educational system today. In 1986 the Lei de Bases do Sistema Educativo (LBSE) (Basic Law of the Education System) received approval. It provided a set of both general and specific principles and goals for the Portuguese system of education. Within this framework, reference is made to environmental education with a particular emphasis upon education for citizenship.

The LBSE regulates the National Curriculum, although individual schools are allowed some autonomy in order to develop local educational projects, according to local contexts. Suitable areas were identified within the curriculum for the development of multidisciplinary activities, in which environmental education is placed not as an individual subject but as the focus of a set of activities, for which the aim is education *for* the environment.

So at all levels of compulsory education, the LBSE has provided an opportunity for the integration of environmental education:

3 Today in Portugal, teaching is divided as follows:

Kindergarten - 3 to 6 years old; situated in buildings next to primary schools or in separate buildings; they have kindergarten educators; they do not have a curriculum but they follow the general principles and aims defined in the Basic Law.

Basic School - (compulsory schooling) is divided into 3 cycles: *1st cycle* - 4 years; global teaching; taught by only one teacher; situated in Primary schools. *2nd cycle* - 2 yrs; taught by several teachers; the curriculum is organised into interdisciplinary areas; situated in Preparatory schools (until the approval of the Basic Law of the Education System compulsory education ended here). *3rd cycle* - 3 yrs; taught by several teachers; the curriculum is organised into disciplines or disciplinary groups; situated in Preparatory and Secondary schools.

Secondary School - *professional*: variable duration; taught by several teachers; organised into disciplines with a strong practical component; situated in professional schools and grants diplomas. *Complementary*: 3 yrs; leads to University or Superior Schools.

Higher Education - 3 years; grants Baccalaureate & with further study grants a Degree; can be State or Private institution.

- *Formal education* - a compulsory period has been set aside in the curriculum, where projects with a global theme can be undertaken, in which 'education *for* the environment' can be developed as a parallel purpose in planning teaching activities. The introduction of *Personal and Social Education* (Educação Social e Pessoal) as a subject within the curriculum, provides further scope for teachers to include environmental education within their work, (Oliveira - Formosinho, 1988). PSE combines aspects of ecological education, consumer education and health education.
- C*omplementary activities* - the opportunity for informal environmental education to take place through clubs and extra-curricular activities is highlighted. Although attendance is not compulsory, it may be possible for teachers to reach a significant number of pupils and students and encourage them to become involved in environmental projects in this way.

A commission was formed by members from the Ministério da Educação (Education Ministry), the INAMB and the Instituto Nacional de Defesa do Consumidor (National Consumer Defence Institute), in 1989. This commission produced a report, "Introdução da Educação Ambiental e do Consumidor no Curriculo" (Introduction of Environmental and Consumer Education in the Curriculum), which suggested that environmental education should be incorporated into curriculum plans for teacher training, that adjustments needed to be made within the infrastructure of the educational system to accommodate environmental education and that environmental education should be encouraged at all teaching levels.

In summary, the education system in Portugal is currently backed by a curriculum and Ministry of Education policies which enable and encourage the development of environmental education in Portuguese schools. It has also begun to focus attention upon both the quality and content, scientific and pedagogical, of teacher training courses for environmental education.

TEACHER TRAINING IN ENVIRONMENTAL EDUCATION

One of the features of the LBSE was that the profile of pre-service teacher training was increased within the education system and that it advised that in-service training should serve to compliment this process and act as a tool for up-dating teachers on new developments.

A revised regulation structure was implemented through the LBSE for teacher training in 1989. The training of teachers (except at *superior* level) was made much more compatible with the general goals of the overall education system. In so doing, consideration was given to the results of experiments conducted in the years leading up to the reform of the education system, while at the same time trying to develop the capacity of teachers to tackle the scientific aspects of environmental issues. The LBSE also describes the role of investigation and innovation as an essential long-term aspect of teacher training.

FEATURES OF THE TEACHER TRAINING SYSTEM

- By law, pre-service teacher training in Portugal should include personal, social and cultural training.
- Pre-service teacher training for kindergarten educators and for basic and secondary school teachers is provided by Escolas Superiores de Educação (Superior Schools of Education) or by Universities.
- Although the general structure of teacher training courses is set down in legislation, the development of curricula is undertaken by the Superior Schools and Universities, who have pedagogical autonomy.
- One of the important features of in-service teacher training is the emphasis placed upon self-appraisal and assessment by teachers.
- In-service training is provided on a decentralised basis.

Priority areas for in-service teacher training are identified each year by the Education Ministry. This training is then provided through one of seven Direcções Gerais de Ensino (General Teaching Departments) (including the Islands of Madeira and Azores) by a number of different types of institutions - Universities or Superior Schools, Training Centres attached or linked to School Associations, and NGO'S. Proposals for training initiatives from any such institutions must match the criteria outlined by the Ministry. Courses generally last from between 15 to 30 hours and are essential for career progression. Currently, high quality training models are given approval by the Conselho Científico-Pedagógico da Formação Contínua (Scientific-Pedagogical Council of In-service Teacher Training), while standards are assessed and set by the Direcções Regionais do Ministério da Educação (Regional Departments of the Education Ministry).

Given this background, what is the situation regarding the provision of training for teachers in environmental education? It is, as has already been mentioned, difficult to assess accurately the position of environmental education in teacher training in Portugal. However, a study conducted by the Instituto de Inovação Educacional, IIE, (Institute of Educational Innovation) in 1993, helps to shed some light on the question.

In order to acquire data about the inclusion of Environmental Education in the curricula of teacher training programmes, the IIE sent a questionnaire to the relevant Universities and Superior Schools of Education. The response rate was greatest from the Superior Schools of Education, and so the discussion that follows relates only to the pre-service teacher training courses or the 'basic education' of teachers, provided by those schools. The following points were noted:

(1) only under exceptional circumstances is environmental education considered as an autonomous subject. More often it links with other subjects, which frequently have compulsory attendance;

(2) most of this type of training is delivered during the second year of such courses;

(3) the average duration of environmental education training for each course is 26 hours;

(4) the topics covered within 'environmental education training' included: climate change, global warming, treatment of solid and liquid detritus, erosion of natural resources, soil quality, mining, water pollution and management, renewable and non-renewable energy, ecosystem dynamics, threats to biodiversity, urban stress, individual and community attitudes towards the environment and whole school approaches to environmental education.

In the same year, 1993, the following areas (See below) were prioritised by the Education Ministry for in-service teacher training:

Table 4.1 1993 priority areas for in-service teacher training

Pedagogical evaluation	Foreign languages
Personal and social training	Language and Portuguese culture
School-grounds*	Methodology and teaching technics*
Technologies for active life	Other teaching domains*
Experimental sciences and their didactics*	

* *Courses on or including scope for environmental education*

The FOCO/INTERGEST Programme of in-service teacher training, organised by the IIE, which ran under the 1993 Ministerial guidelines, found that environmental education amounted for only 3% of in-service courses.

Each course lasted at least 25 hours. The most frequent themes were: education for the environment, methods for study of the physical and social environment, local heritage (natural and/or cultural), health education, intercultural education, integration of environmental education in the curriculum and environment sciences.

INAMB-IPAMB COURSES: 1988 - 1992

Throughout the period 1988-1992, the INAMB-IPAMB organised a number of courses designed to meet rising requests from teachers and schools which were made either by individuals themselves or through co-ordinating organisations such as the General Departments from the Education Ministry, City Halls and Centres of Pedagogical Support, (see Table 4.2).

Table 4.2 Training Courses for primary school and secondary school teachers

Year	1988	1989	1990	1991	1992
Total Courses	23	15	35	36	29
Total Teachers	1000	500[4]	1700	1275	873

Most teacher training courses were designed for teachers of 1st and 2nd cycle of Basic School. Educators at kindergarten level received less of such training than any other group. A general course on the aims and purpose of environmental education was also organised in this period and proved a great success being attended by 333 teachers.

Along side these training courses, the INAMB-IPAMB also financed 37 environmental education projects. These projects were developed in Primary and Secondary Schools, between 1991 and 1992. This programme continued into 1992-1993 with the involvement of 32 new schools.

4 There was a change in the teacher enrolment system in 1989.

Representatives from the IPAMB also work with the Ministry of Education to support a number of environmental education projects lasting one year or more within a selection of schools. Project guidelines, rules and deadlines and set by a team from both the Ministry and the IPAMB.

Due to financial constraints and the fact that more and more schools apply to be selected each year, specific criteria are laid down for the selection team to assist their choice.

With the financing they get, the selected schools buy material and pay for students transport, when field visits are required. Local Authorities often provide extra assistance in terms of transport particularly for schools dealing with very young pupils.

Field visits usually take no more than one day. Within the formal education structure, it is not possible to take any longer. Organisations such as the Instituto Nacional da Juventude (Youth National Institute), Boy Scouts and on the Associations for the Defence of Environment who arrange camping trips, pedagogical itineraries and, less frequently, visits of more than one day to Nature Resources Centres, generally tend to take responsibility for longer field trips.

Normally the project concludes with an exhibition of the work conducted by the students. These exhibitions are usually open to the community.

The results of the projects are later evaluated by the Financier Institute, which analyses a report of pedagogical activities and the methodological processes that were used throughout the projects. They also review each school's budget and justifications for expenditure. Information on a variety of the activities carried out by individual schools is disseminated via the free, monthly newsletter of the IPAMB.

The IPAMB also promotes a literature contest entitled "The Environment in Youth Literature" and organises various seminars and national meetings on environmental education, in which all teachers can participate.

CEFOPE[5] AND ENVIRONMENTAL EDUCATION

Although the decentralised nature of teacher training in Portugal provides an answer to local needs and demands, it also acts as a barrier to the development of a comprehensive picture of courses and activities taking place countrywide. For this reason it seems useful to describe, as an example, some detailed information about the teacher training programmes for environmental education run by Centro de Formação de Professores e Educadores de Infancia, (CEFOPE), at the University of Minho.

Pre-Service Training

CEFOPE provides pre-service teacher training courses for both kindergarten educators and primary school teachers. Environmental education is built into the curriculum plans for kindergarten educators. Apart from disciplines where environmental education is 'implied', such as Social Sciences, Natural Sciences, Social Education, Health Education, there are two further (alternative) disciplines called Protecção do Património Cultural (Cultural Patrimony Protection) and Educação Ambiental (Environmental Education) included within their curriculum. Within the primary school curriculum environmental education is not included as an expressed discipline although it is 'implied' through the disciplines of Ciências Naturais (Nature Sciences), Ciências Sociais (Social Sciences) and Educação Social (Social Education). The results of teacher training in environmental education can be seen through projects developed with children from nursery schools and from primary schools, in the disciplines of Prática Pedagógica[6] (Pedagogical Practice), and also through the exhibition of work completed by students in the schools where these projects are developed.

In-Service Training

CEFOPE also offers courses of specialisation in superior studies for kindergarten educators and for 1st cycle of Basic School, which have been in existence since 1991/1992 and aim to increase the professional training of teachers. These courses take place after working hours, last two years and course sessions have compulsory attendance for those who enrol. The curriculum plans have disciplines that are linked to environmental education, such as the Educação Pessoal e Social (Personal and Social Education) - which is common to several courses - and specific disciplines on environmental education such as Educação Comunitária Ambiental (Environmental Community Education), Metodologia do Estudo do Meio Local (Methodology of Local Environment Study) and Educação Ambiental (Environmental Education).

Each course is organised into four semesters. The last deals exclusively with the development and execution of an individual research project, chosen

5 CEFOPE is a training centre for teachers of 1st (and in the future) 2nd, cycle of Basic School and for Kindergarten educators. CEFOPE is associated with the University do Minho in Braga.

6 This is a teaching model, which examines the needs of theoretical-practical integration, (Formosinho, 1985, 1986).

from among a proposed set of themes. The methodological characteristics of each research project are different, although a great number of students choose action-research. Several teacher-students chose to work with environmental education and some conduct action-research projects in the schools where they work. Using theoretical information from the previous semester, the teacher-students design a curriculum project for their schools. The project focus itself is usually a small, real environmental problem, relevant to the school and community, while at the same time providing scope for the organisation of topic plans and educational outcomes.

The main purpose is to solve real environmental problems and to educate children for the environment, using small scale projects. The school is considered as a doorway to the local community. The teacher-students first try to engage the attention of parents and then of the local authorities, environmental associations and environmental experts. All actions and activities undertaken during the project undergo continuous evaluation. Each project is completed by a final report which must describe the activities attempted and those completed as well as the pedagogical-didactic processes followed, teacher-students must discuss both the main sources of potential development and the main obstacles encountered during the course of their projects. The reports are kept by the University library.

CEFOPE also runs a Research Centre in connection with the Education Sciences Department. In 1993 and 1994 a European research project was developed, bringing together four European Universities from (Denmark, France, Portugal and United Kingdom). This project and the subsequent report, was called "Children as Catalysts of Environmental Change"[7]. This project gathered kindergarten educators and other teachers, who after attending training courses worked on environmental education projects with their pupils.

Courses on Environmental Training and Methodology in Project Development were delivered as part of this initiative and a resource centre equipped with didactic materials and relevant literature was created. This Centre had scientific support from 3 experts in Environmental issues, throughout the initial stages of the project. Two one-day conferences, dedicated to environmental and environmental education themes, were designed by representatives from the University - both students (teachers) and lecturers -, INAMB-IPAMB, a local Museum and representatives from an environment defence association.

During the coming school year (1995 -1996) CEFOPE will have developed a new course, designed to provide teacher training in

7 European Commission: DG XII / D5, Research on Economic and Social Aspects of the Environment (SEER1).

environmental education at 2nd cycle of Basic School and at superior studies level.

The activities carried out by CEFOPE detailed above serve as more than an interesting collection of references. They serve to illustrate the fact that in-service teacher training for environmental education developed by the Education Ministry through the FOCO/INTERGEST Programme, forms only a small part (3%), of such training at local level countrywide. It also serves to highlight the very real and concrete demand for environmental education from both pre- and in-service teachers in Portugal.

ACKNOWLEDGEMENTS

I would like to thank: Professor Patricia Fontes for the opportunity to write this chapter; Professor Bártolo P. Campos, IIE President, for the permission to use IIE data; Dr Isabel Raposo from the INAMB-IPAMB; and my friend Júlia Formosinho for her personal support.

REFERENCES

Evangelista, J. 1992. *Razão e Porvir da Educação Ambiental* Lisboa.

Fernandes, J. A. 1983. *Manual de Educação Ambiental* Lisboa: SEA: Comissão Nacional do Ambiente.

Fontes, P. & Máximo, L. 1994. National Case Studies: Portugal. In: *Children as Catalysts of Environmental Change: Final report.* European Commission Directorate General for Science Research and Development. Joint Research Centre: DGXII/D-5 Research on Economic and Social Aspects of the Environment (SEER I): Contract Nº EV5V-CT92-0157.

Formosinho, J. 1985. *Da formação inicial como ponto de partida à formação contínua como ponto de chegada.* Universidade do Minho Braga.

Formosinho, J. 1986. *Fundamentos da formulação e gestão curricular da formação inicial de educadores de infância e professores primários.* Universidade do Minho Braga.

Oliveira-Formosinho, J. 1988. A Educação Pessoal e Social na Lei de Bases do Sistema Educativo e na Reforma educativa. *Parecer elaborado para a Comissão de Reforma do Sistema Educativo.* Braga CIFOP-UM.

CHAPTER 5

ENVIRONMENTAL EDUCATION IN TEACHER EDUCATION IN FINLAND

Sinikka Ojanen & Mauri Åhlberg

INTRODUCTION

A unique feature of Finnish teacher education is that a master's degree is required from all comprehensive school teachers including those working at primary level. Primary school teachers become Masters in education and subject teachers become masters of their major teaching subjects. Given that there should be no definitive line between environmental and other types of learning, this situation provides distinct possibilities for teachers to undertake teaching and research in environmental education alongside their other work.

In-depth and meaningful learning is that which is accepted as a part of one's own life. Environmental education should be inquiring, problem-centered, multidisciplinary, holistic and it should also involve a personal relationship to the questions of nature and human construction with respect to environmental quality. Man's own environment of perception and experience, his values and beliefs, the psychological environment and the cultural entity, too, are central and crucially significant features. On the other hand it is important to realise that environmental education is not a shift into dogmatic, vague and intuitive holism. The scientific view of knowledge implies that knowledge can always be examined and tested. Similarly, practice will partly function as a test, if the results or consequences are studied during or after the action itself. Action as such does not guarantee learning.

Environmental education, at it's best, surveys the dynamics of man and his relationship towards nature and the environment and recognises and supports this reciprocal relationship. This should be a continuous process, in which people become aware of their environment and begin to reflect upon the consequences of their decisions. Of great significance to this process is the way in which society manages, for example, the psychological working environment and the safety of the environment in which our children and grandchildren will grow up. Equally important is the definition of human welfare or health and the type of ethical significance attached to them.

Not merely ecology

As a concept environmental education is far from unambiguous, instead it is a wide, multidisciplinary entity of problems. It originates from environmental protection but during a relatively short history the content and character have been developed in a more and more holistic way. It combines in an in-depth manner the societal and the scientific knowledge of man. The 1990s have seen a clear shift an extension of emphasis. This so-called 'extended' environmental education is no longer mere ecology. It is understood as a multidimensional entity: **psychological** environment (personality); Interpersonal **social** environment, studied by interactional sciences, for example social psychology; **man-made, cultural** environment and so on. The human day-to-day environment consists of all the above in the way a person simultaneously perceives it.

Environmental education is also ethics education

A precondition to human psychological growth is not only the ethical reflection of oneself but also of the environment and man's relationship with the environment. Environmental education is a form of education which aims at affecting human values and principles. It is based on environmental values, to which people are committed and which influence their feelings and in this way seeks to create a basis for attitudes and behavior. A basic characteristic or cornerstone of humanity for each person - and especially for an educator - should become the raising of human ethical awareness. It means a shift of emphasis towards the importance of human psychological growth from the present material ideal.

NATIONAL AND INTERNATIONAL RESEARCH IN FINLAND

Characteristically, environmental education research is applied, multidisciplinary and societal in nature. In practice this means that such research mainly originates from practical life problems, for which solutions are attempted by applying existing knowledge to new experience based knowledge.

A problem in the instruction of environmental education at both schools and universities has been the fact that the teaching of values has been doomed to fail. Ready-made knowledge is not easily converted into subjective skills. Such a process requires both interdisciplinary co-operation and a new view of educational possibilities.

In the autumn of 1994 the Finnish Ministry of the Environment organised the first ever national environmental education researchers' meeting. Most of the participants were teacher educators. At present teacher educators are actually responsible for the most part of the research in this field in Finland. Teachers graduating from Finnish universities also receive research training during initial education.

It is of great importance that the field of environmental education aspires for specifically in-depth and meaningful learning. Reflection has been promoted by using for example, experiential learning, concept-mapping and V diagrams. Environmental research has traditionally dealt mainly with minor, natural scientific phenomena, whereas educational research takes account of the learning situation itself by observing the learning process of both pupil and teacher. Similarly, it is important to be aware of what happens elsewhere, thus links and networking are vital.

To this end, the Nordic Countries have an active interschool research and exchange programme. Finnish schools also take part in the Scandinavian MUVIN research project, where the solving of environmental conflicts plays a central role. At present the project also includes small-scale educational research. The project, which started in 1991, involves 250 schools, several departments of teacher education and researchers. There is a Scandinavian data-base under way, the individual schools' ventures have been listed and published; mutual communication is taken care of by the Norden-Muvin bulletin. The Nordic Council of Ministers financially supports the teacher and pupil exchange programme. The first phase of the project ended with an environmental education event and competition organised by Norway. The Competition was won by The Kalevankangas junior and senior secondary school in Mikkeli, Finland. The MUVIN project also offers a teacher education course likewise funded by the Council of Ministers.

ENVIRONMENTAL EDUCATION IN TEACHER EDUCATION IN FINLAND ON THE BASIS OF AN EMPIRICAL STUDY

In spring 1995 the Savonlinna Department of Teacher Education (SDTE) surveyed and interviewed the representatives of all Finnish departments of teacher education about their environmental education programmes. As a result of this study it was concluded that the teaching of environmental education in the various departments (12 in all) is undertaken as part of the individual profile of each institution, but that there are some coherent packages to be seen - basic, specialist and optional studies. The teaching of environmental education is a generative, fluctuating process, which includes both starting and terminating of compulsory courses. An increase in the supply is a clearly visible feature as is interfaculty co-operation. Details of the different departments of teacher education (DTEs) will now be presented in alphabetical order.

Helsinki Department of Teacher Education

Helsinki DTE offers a 15 credit package of environmentally-oriented science education (not synonymous with environmental education) as specialising studies and has twice before offered optional studies (4 credits): Ethics in environmental education, Introduction, Method, Analysis of problems. Some 30 classroom student teachers have taken part in the course and positive feedback has been received. Environmental Education (1 credit) has also been taught as a compulsory unit in the past. The course consists of 20 hours of lectures and 20 hours of self-directed work. The Department of Forestry also offers student teachers an environmental sciences package of 15 credits.

Tampere University, the Hameenlinna Department of Teacher Education

Tampere also offers a 15 credit environmental sciences package as specialising studies (including a unit on didactics), 20 student classteachers acted as organisers for 1995. Basic studies include 5 credits worth of sciences, one credit of which is environmental education. The institution actively aspires for compulsory environmental education courses.

Joensuu University, The Joensuu Department of Teacher Education

Attempts have been made to integrate environmental education into different subjects. Lectures and demonstrations in environmental education (teaching and learning) 1-2 credits are offered as options for the classroom student teachers. Students from the Open University also participate in this section. A 15 credit intra-faculty environmental studies package has been designed. However, only a few (6) classroom student teachers have taken part in this education

Joensuu University, the Savonlinna Department of Teacher Education

Basic studies contain 5 credits of optional studies in environmental education:

(1) Ecology and Environmental Education;
(2) Environmental Philosophy and Ethics;
(3) Environmental Knowledge.

These courses have had some 60 participants. Starting from autumn 1995, basic studies in Savonlinna will include 1-2 credits of compulsory environmental education.

The Jyvaskyla Department of Teacher Education

Environmental education is offered in specialist studies (15 credits) and there are, in addition, optional courses available.

The Oulu Department of Teacher Education

The education of classroom teachers has no compulsory units in environmental education. The institution is currently aiming at adding such study into the classroom teacher programme. The programme had at one time 2 credits worth of the above-mentioned study. A future module, of 15 credits, will be carried out in co-operation with an Open University group for lectures and then student teachers will attend their own seminar group.

The Rovaniemi Department of Teacher Education

Rovaniemi offers the following compulsory basic studies in environmental education: The development of child and youth scientific worldview (biology and geography; science, natural sciences, civics) - integrative studies 2 credits. Optional, integrative studies including Man and His Environment (2 credits) and specialist studies:

(1) Early childhood education: science;

(2) Biology/geography: Man and the Environment on Earth; Environmental Protection, Man and the Environment in Finland, Biological Resources, Natural Geography Environmental Education have been integrated into biology and geography.

The Turku Department of Teacher Education

The curriculum here does not include a joint environmental education section at present. There was, in 1992, a compulsory course of 2 credits in environmental education. The course had 150 participants. Since then there has been no funding for such education.

During the years 1992-95 there has been an integrative basic studies environmental education project, in which 20% of each year's students have participated. In the pro-gradu thesis seminar level, between 4-6 students, and on an average 2 working pairs, annually choose topics relating to environmental education.

Åbo Akademi, Vaasa Department of Teacher Education

"Miljöfostran", a multidisciplinary, environmental education is integrated into basic subjects, mainly biology, geography, art education and early childhood education. Elective studies include 5 credits of environmental education. Currently, a 15 credit study package of environmental specialisation studies is under consideration.

THE TRAINING OF TUTORS - TUTORS AS GUIDES

In Finland the promotion of environmental education has been given due attention by numerous governmental documents and recommendations. The Ministry of Education has funded a project: *The Environmental Education Programme for Teacher Educators*. This project involved training in all 12 teacher education units in 9 universities. Teacher educators were identified as a critical target group, because they were thought to be central in spreading and implementing new practices. All these teacher educators are working in universities and university practice schools, where student teachers undertake their training.

An environmental education course for practicing teacher educators in Finland was organised as an initiative of the Ministry of Education. A 'tutor training course' was planned and attended by 1-3 persons from each of the 12 university teacher education units. The tutor course started in December 1992 and 17 educators completed it within a year. They participated in a unique learning system, studying environmental philosophy and ethics, educational theory and teaching methods relevant to environmental education. They also studied environmental sciences. An important part of the programme was training in distance education, tutoring and course planning. The trained tutors then started to lead study groups in their departments to train their colleagues to implement environmental education practices as part of a teacher education curriculum. By the end of 1994, approximately one year after it started, about 200 teacher educators had begun this programme. The expected study period for each person is one year, but may vary from six months to two years. The goal is to educate the majority of teacher educators. The course is free for the participants and many have included it as a part of their post-graduate studies. The focus is upon personal development and curriculum development. The project will end in 1996.

A simplified and modified version of Hungerford & Volk's (1990) model on the development of environmentally responsible behaviour was chosen as a theoretical framework, which has not been accepted by all environmental educators due to, for example, the linear behaviorist nature and wide empirical basis. This model has, however, great heuristic power in directing educational practices and research. It has also worked well as an advance organiser in different parts of the course and helped to combine the various aspects into a more or less coherent framework.

Hungerford & Volk's model was complemented by critical reflection ability so that environmental education became a cultural critique. This ability is needed to make individuals conscious of values which harm and destroy nature and those that support life. We cannot change the social structure of our beliefs, values and actions, if we are not able to reflect on them. Environmental issues are so complex that an ability to understand and evaluate them demands the ability of critical reflection. The course programme was developed to include enough 'space' to evoke sensitivity to nature, promote reflection and learn different kinds of knowledge and skills in using environmental action strategies.

The University of Jyvaskyla was chosen as the focal point for this project due to the geographical location and strong tradition in educational studies. The aim of the planning stage was to create a sound theoretical basis through literature review in order to choose an appropriate implementation strategy.

The method selected was distance education. This included some locally arranged contact teaching. The local study groups were lead by previously trained tutors. Individuals attending the course applied teaching practices that were novel to them and produced new types of model lessons. They also reflected upon their own personal environmental philosophy and deepened their environmental knowledge. The preliminary findings of the evaluation research indicate high quality learning experiences. It is suggested that this model may be useful for the design and implementation of a wider-scale environmental education in-service teacher training course.

Programme Content

The contents of the course are divided into three modules. They are:

(1) environmental philosophy and ethics;
(2) pedagogical theory and teaching methods relevant to environmental education;
(3) environmental sciences.

Initially, it was thought that environmental ethics would play but a small role within the overall framework. However, as the course developed, it was decided that a thorough reflection of individual environmental philosophies is vital component. Many tutors thought that this was the most rewarding part of their own course.

Teaching methods were examined through workshop demonstrations and literature review. Emphasis was placed upon value education and issue investigation strategies. In this way a holistic model for environmental education has been created using confluent education as a basis. Through this model, teaching strategies have been developed which aim to reveal the implicit social meanings of the environment and to examine symbolic meanings which clarify why some problems become hot social issues while many others do not arouse public interest.

The third module in the curriculum is 'environmental knowledge'. The contents of this section were chosen individually to match the needs of teachers from different subject backgrounds. A chemistry teacher needs a different body of knowledge than, for example, a history or a crafts teacher. For this reason, a multisubject issue-oriented approach was chosen.

Tutors (students) were set several learning tasks or assignments based upon the three modules. Each assignment was preceded by a reading 'package' of selected articles and essays. Many students have used these assignments to write articles and/or to continue their formal studies. Each assignment is evaluated by a course tutor and a selection are sent to the central organising body of the course for monitoring purposes.

MIKKELI PROVINCE ENVIRONMENTAL PROJECT

In the heart of the Saimaa Lake District, Mikkeli ecocounty/province has sought to developed the ideas and principles of environmental education according to the spirit of *Agenda 21* - the UNCED document. Mikkeli Province has long aimed for a sustainable way of life and is currently a demonstration area for environmental education. It has already produced both nationally and internationally valuable experience for the research and development of environmental education.

In March 1994, the *Finnish Sustainable Development Committee* selected the Environmental Education Project in Mikkeli Province to be the first official Finnish UNCED (United Nations Conference on Environment and Development, Rio de Janeiro, June 3-14, 1992) project for environmental education.

This project was based upon an agreement of co-operation with " a school" for long-distance runners. As a result, the notion of sustainable development was put into practice at the 1994 European Championships in Athletics in Helsinki. Furthermore, the agreement also included the environmental education of trainers for 15,000 young athletes.

Mikkeli Province is developing a remarkable network of research centers and specialised museums which serves the needs of environmental education very well. The *Rantasalmi Lakeland Center*, the *Finnish Forest Museum and Information Center, Lusto* were opened in the summer 1994. For example the *Beekeeping Museum* in Juva, *the Lake Fishing Museum* in Kerimaki and the *Provincial Museum* in Savonlinna organise exhibitions with environmental themes. The *Lakeland Center* has a versatile education and development programme which has resulted in more extensive resident outdoor education activity, development of environmental school network and co-operation with the information center of the *Linnansaari Nature Reserve*. The World Wide Fund for Nature (WWF) supports the development of the Lakeland Center. Mikkeli, on the other hand, has the *East Finland Green Information Center*.

UNCED provides Mikkeli Province with an opportunity to participate in the UN's environmental programmes. Furthermore, it opens up new prospects for versatile co-operation and the implementation of practical applications of environmental education internationally. The objectives of the Mikkeli Province Environmental Education Project for 1995 are as follows:

- The development of an environmental school network
- Network university and research centers as an international model within the UNCED Programme
- Environmental education courses
- Summer terms for vocational schools
- Outdoor education
- Research in environmental education

The main areas of emphasis within the plan are:

(1) a sustainable way of life;
(2) environmentally friendly technology;
(3) ecologically beneficial use of land.

The researchers of this project work together in order to utilise the available material in as versatile a manner as possible and in order to maintain the proportion and representative balance of the material. The planning and writing of nine dissertations which deal with environmental education in primary and secondary schools and in the vocational areas of health care and textiles, has already started within the project.

The dissertations have been integrated into a coherent research programme called "A sustainable way of life through environmental education", which is being developed by the Academy of Finland and the European Union. According to this project there is presently a need for two kinds of research:

(1) Basic research, clarifying the nature of environmental education as interaction; what teachers and pupils learn, think and feel, and how they react to environmental education. The aim of basic research is to describe a situation and the way it changes. This knowledge is a precondition for the development of environmental education and for the assessment of its effects. To find answers for environmental problems requires creativity in problem solving. Therefore, it is important to find out in what kinds of schools and circumstances the most, as well as the fewest, fruitful new ideas are produced. There are, of course, many other issues to be handled within basic research.

(2) Applied research, the results of which are used directly to improve environmental education. One of the most effective forms of applied research is so-called action research. In this type of research the researcher takes an active part in promoting the development of environmental education. In practice, also the teachers and pupils are guided to acquire knowledge in a manner similar to research work. In this way, they are able to analyse the complex problems of day-to-day life. The aim is to teach them to develop the solution for usually complex environmental problems."

SDTE - THE UNIVERSITY OF MIKKELI PROVINCE

SDTE, a filial department of the University of Joensuu, is located about 150 km from the main campus of Joensuu and shares a campus with the Joensuu University's Savonlinna Practice School (an experimental school) and the School of Translation Studies. The Open University and the Summer University also operate from Savonlinna.

The environment of Savonlinna is in many respects, exceptional, being internationally renowned for beautiful and peacefully natural surroundings. The buildings and technical equipment of SDTE are modern and very well suited for the purposes of teacher education and research.

THE CURRICULUM OF SDTE

Environmental education, as such, is taught at SDTE but is also integrated into other subjects like art education. Insight and experience and their meaning are given specific emphasis. There are separate courses about environmental aesthetics, which focus on the observation of nature and the built environment as well as the aesthetic form and quality of the material environment. Environmental education is integrated into education courses, for example, courses on learning instruction, curriculum and research methodology. A crucial starting point in terms of 'values' is that of sustainable development. The education of classroom teachers also includes both a subject science and ecology based course in environmental education.

Since the SDTE programme includes research-based education at doctorate level, which helps to combine both theory and practice of environmental education, the tutors involved with the Mikkeli ecocounty sponsored teachers' in-service environmental education project, seminars and research activities have thus far largely been SDTE professors.

SDTE has actively sought to develop involvement in the design and implementation of regional policy as well as of national teacher education specifically in particular areas of expertise such as environmental and art education. SDTE has also produced learning a range of learning and practical resource material, like for example, an environmental education handbook.

International co-operation and networking

Students from SDTE regularly go abroad to study either as part of an exchange or for further study, and environmental education is often the field of study chosen. For these purposes SDTE is a member of the Erasmus project and also has links with other organisations working in the fields of teacher training and environmental education such as - ATEE, ATE (in US), AERA, EARLI, ECHA, CARN, ISATT. Current environmental education research topics include: student orientation, reflective teaching and teacher education, significant learning experiences, pedagogical communication, student empowerment and environmental education, hypermedia in language, learning picture reading skills.

CONCLUSION

Teachers cannot be offered a ready-made model for either environmental education or for any other education. Generalisations and universal rules easily prevent teachers from taking into account each individual context. It is important, however, to help teachers find and preserve a creative attitude towards their work, so that they can constantly develop within their profession. This will help to encourage a sense of meaning in their work.

Although international evaluations concerning the environmental education policy in Finland have been quite flattering, we believe that there still is a long way to go. Environmental education should change from special projects to everyday practice in an integrated way.

ACKNOWLEDGEMENTS

The following persons have contributed in preparation of this chapter:

Project Manager Pekka Hynninen, BSc, Rantasalmi Institute of Environmental Education; Dosent Dr. Markku Kapyla, University of Jyvaskyla; Researcher Riitta Wahlstrom, Phil.lit., University of Jyvaskyla.

REFERENCES

Hungerford, H. R. & Volk, T. L. 1990. Changing Learner Behavior Through Environmental Education. *Journal of Environmental Education Spring, 21* (3) pp 8 -21.

BIBLIOGRAPHY

Hynninen, P. & Ilmavirta, V. 1995. *Mikkeli County Environmental Education in Mikkeli Ecocounty: The realisation of the UNCED-process as a network* Unpublished.

Käpylä, M. 1994. Yhdentävä Kasvatus Ympäristökasvatuksen holistisena viitekehyksenä (Confluent education as a holistic framework for environmental education). In: *Käpylä, M & Wahlström, R. (Eds.) Ympäristökasvatuksen mene elmäopas (A methods guide for environmental education). Jyväskylän yliopiston täydennyskoulutuskeskuksen tutkimuksia ja selvityksiä 17*, pp 10-19 University of Jyväskylä, Continuing Education Centre.

Käpylä, M. & Wahlström, R. (Eds.) 1994. *Ympäristökasvatuksen menetelmäopas (A methods guide for environmental education) Jyväskylän yliopiston tädennyskoulutuskeskuksen tutkimuksia ja selvityksiä 17* University of Jyväskylä, Continuing Education Centre.

Ojanen, S. 1995. Mielekkään oppimisen malli ympäristökasvatuksen oppettamiseen. Ihmisekologian teoreettista tarkastelua. Teoksessa (tiom. Ojanen & Rikkinen): Opettaja ympäristökasvattajana. (The model for meaningful learning in teaching environmental education. Theoretical study of human ecology). In: Ojanen, S. & Rikkinen, *Teacher as Environmental Educator* Helsinki: ENSOY.

CHAPTER 6

PROFESSIONAL DEVELOPMENT OF TEACHERS IN ENVIRONMENTAL EDUCATION: ENGLAND AND WALES

Chris Gayford

CURRENT TRENDS AND INFLUENCES

To appreciate the thinking behind the array of materials produced to support the professional development of teachers in environmental education in recent years in England, Wales and Northern Ireland by non-government organisations (NGOs), such as the Council for Environmental Education (CEE), several important influences should be considered. All of these have been the subject of recent rapid change. Firstly, the nature and purpose of environmental education has undergone a major shift of emphasis in recent years. Secondly, the general context in which teachers are currently expected to work has endured massive changes and finally, the role of voluntary organisations themselves has altered, particularly in relation to the way in which they are now expected to function at local, national and international levels.

Following the United Nations Conference on the Environment and Development (UNCED) in Rio de Janeiro (United Nations, *Agenda 21,* 1992), popular ideas of environmentalism have modified significantly. Prior to this time there had been a gradual evolution in thinking among the more informed environmentalists, so that ideas relating to development and long term sustainability were accepted as central to any concept of environmental conservation. This thinking had not, until recently, begun to penetrate education relating to the environment in schools. Following on from UNCED, sustainability, and all that this implied for environmental education, became part of a more widely appreciated agenda. The UK (DoE, 1994), along with the other signatory governments, recognised the important contribution of environmental education as part of a planned programme for achieving the aims of *Agenda 21* where environmental education was identified as an essential cross-cutting element to the school curriculum (United Nations, 1992). To help achieve these aims it was clear that schools would need a considerable amount of support in many ways, not least of which is through the professional development of teachers. One of the main problems for practising teachers is to feel confident in addressing the range of disciplines that can contribute to an holistic appreciation of environmental issues (Gayford, 1991). Part of the problem is the reductionist way that knowledge is currently divided up in the curriculum through the dominance of traditional subjects. This works against the development of the more synthetic, and perhaps eclectic, thinking that is required in environmental education (Sterling, 1992; Smyth, 1995). The issues that this raises for schools are significant and surround debates about both the nature and purpose of environmental education, as well as its place in the curriculum. This in turn raises questions about the most appropriate ways to support teachers.

The changing context in which schools function following the Education Reform Act of 1988 is another, and equally important aspect relating to the changed circumstances in which in-service training is being developed. This has manifested itself in a number of ways. Perhaps the most obvious are:

(1) The enormous impact of the National Curriculum, which reinforces approaches to the curriculum through strong emphasis on the teaching and assessment of traditional subjects.

(2) The introduction of local financial management of schools (LMS), whereby each school has responsibility for its own budget, including in-service training. This has had considerable implications for the allocation of time and resources for teachers to acquire the necessary knowledge and skills to become effective environmental educators. The new arrangements for INSET through the GEST scheme (Grants for Education Support and Training); in which the rules governing the way in which funds for this purpose will be allocated have been radically revised. Part of this relates to identified national and local priorities (DFE, 1992), which tend to focus on support for the delivery of the National Curriculum and school management issues. The general demise of the local education authorities (LEAs) has usually had a marked effect on the access that teachers have to INSET in environmental education (Hopkinson, 1991). Until recently LEAs have usually provided a centralised service which has often included

advisory personnel who have organised a wide-ranging programme of INSET activities, including those for environmental education.

(3) The recent revision of the National Curriculum by Sir Ron Dearing (Schools Curriculum and Assessment Authority, 1994) was an effort to reduce the amount of over-prescription and over-assessment that was felt to exist. In retrospect this has done little to further the cause of environmental education in schools, since the term 'cross-curricular theme, to which environmental education along with other important educational areas was most associated, has been quietly dropped. One, perhaps hopeful, result of the revision has been the creation of a significant amount of additional discretionary time for schools to use as they wish. Consequently it is conceivable that they could use part of this non-prescribed time for environmental education if they chose to do so; however, there is no specific requirement for schools to include environmental education in the revised curriculum. Therefore with the other pressures that exist, particularly from assessment and inspection (see also The Office of Standards in Education, 1993) there are few incentives for schools to increase the amount of environmental education being done.

Despite these apparently negative influences on the amount of environmental education in schools, recent surveys have shown that the great majority of schools, both in the primary and secondary phases, feel that this is an aspect of education which is of fundamental importance (for example Saunders *et al* 1995). They therefore often make it clear that they wish to continue with environmental education as part of their normal school activities and will seek ways of implementing this within existing frameworks.

There are mixed messages from central government making it difficult for schools to respond in a coherent way. The DoE on the one hand recognise the important contribution that schools should be making towards more sustainable living and the Department of Education and Employment (DEE) on the other hand have been unwilling to give emphasis to the environment in their overall strategies for educational change. One factor that has emerged, and should be recognised as a significant area for developing support for teachers in environmental education, has been the increasing importance of the voluntary organisations (NGOs) in recent years. Their growing professionalism, often accompanied by the appointment of well qualified education officers, has put them in a favourable position to provide support for teachers and schools in a wide variety of ways. So much so that many agencies, including statutory ones, are looking towards them to make an important contribution in this area (see for example the First Report of the British Government Panel on Sustainable Development, DoE, 1995):

> *"Education and training are crucial to the achievement of sustainable development ... [this] applies across the ... voluntary, public and the private sectors."* (DoE, 1994, para 32.12)

Part of the problem for all providers of INSET has been to give an appropriate response to what has often been a general request for help from teachers and schools in this area of education and to identify their specific needs. An important aspect of the work by the CEE, for example, has focused on the nature of these needs and then seeking to address them. The fact that the CEE is a membership organisation which works collaboratively with a wide range of environmental organisations, each of which is able to provide specialised expertise, is a factor that enhances their combined capability of responding to the needs of schools. The Council, through its considerable number of contacts is also able to network with many individual experts and in this way they can, either put the appropriate people in touch with those with a particular training need, or identify suitable people to co-operate in projects designed to provide support for larger numbers of users. Ultimately the emphasis has been on a broad-based approach to helping schools which can be best described using the general term *support*, an important part of which is the provision of in-service training.

INITIATIVES TO SUPPORT ENVIRONMENTAL EDUCATION IN SCHOOLS

These will be discussed from the perspective of the contribution made by the CEE in this field. One of the recent projects undertaken by the CEE is the *Environmental Education Information Needs Project*, funded by the DoE. The remit for the initiative was based upon a Government White Paper, *This Common Inheritance* (DoE, 1991), which acknowledged the need for work of this nature. This particular initiative can been viewed as part of an even broader strategy of support for environmental education, both in schools and in non-formal contexts. One of its main purposes has been the identification of the information needs of teachers (Matthews and Stephens, 1992). Here it was considered essential that in order to begin to address training needs it is first necessary to carry out a needs assessment among the client group in order to identify the type of support that teachers feel they require. The overall purposes of the project in relation to schools can be summarised as the identification of:

(a) The needs of teachers in terms of environmental education materials and resources.

(b) The best means of provision and distribution of environmental education resources.

(c) The effectiveness of sources of information about environmental education resources and materials.

The particular aims were to:

(1) Increase awareness of, and access to, information services.
(2) Encourage discerning use and evaluation of information by users.
(3) Improve and extend the distribution networks.
(4) Advise those who provide information on the needs of users.
(5) Establish ways of keeping information up-to-date and relevant.
(6) Recommend further initiatives.
(after CEE, 1992)

The survey provided a great deal of information which is important for teachers as well as those who provide the support that they think teachers require. It gave baseline data on which to build and which should inform the ways in which the informational side of the professional development of teachers is approached in the future.

From the results of this survey a number of valuable findings were made. These drew particular attention to the importance of each school having on their establishment a member of staff with responsibility for co-ordinating the environmental education of the school. The co-ordinator would assume responsibility for this aspect of the school's function and provide a first point of reference for incoming information, materials and resources. They would also be concerned with curriculum management and a focus for training opportunities. Also where co-ordinators exist, it was felt important that they should be identified and their role assessed, so that examples of good practice and guidelines for the role could be produced and distributed to interested organisations and individuals. In this way a training dimension was seen as an essential part of ensuring that good practice was maintained.

The findings from the study also showed that there was a need for resources that were specifically targeted. Particular reference was made to the fact that these should be targeted towards National Curriculum Attainment Targets and Key Stages and should be evaluated for this purpose. Teachers did not feel that, contrary to general supposition, there was a real deficiency of resources, indeed it was often felt that there was much wasteful duplication. There was in their view a lack of co-ordination among the providers of information, which confirmed some of the original thinking behind the project. They stated that there was a bewildering array of materials and they found it difficult to know what was available, from where it could be obtained and they needed some kind of independent assessment of quality, with teachers themselves being involved in the assessments. Further, there was a need for manuals of environmental education information. The types of information seen as useful were recommended materials , examples of good practice, guidelines on policy, funding sources, lists of contacts and co-ordinators. Manuals should be in a form that could be easily updated and specific local information added. Co-ordinated mailings of environmental education materials within local authorities using *`green envelope'* schemes were seen as a useful method of efficient distribution. The use of information technology was thought to have considerable potential, but currently it is used by only a few enthusiasts. Databases linked to networks, such as GreenNet, are likely to become more widely used as schools develop the capabilities of CD-ROM; but as yet it was not possible to exploit these vehicles for up-dating the great majority of teachers.

Professional training was identified as a vital means of providing support for teachers. INSET of two types is required; one of which is concerned with content and the other with methodology. It was felt that there should also be an environmental education input into all initial training courses for all students. Examples of good practice, wherever it occurred in environmental education, should be disseminated widely. These should include a variety of examples such as lesson plans, curriculum management approaches and ways of evaluating practice. Special emphasis was given by many teachers to materials that had a local perspective. However, it was recognised that awareness of sources of local information, such as local authorities, libraries and local voluntary groups, is poor and requires improvement. Teachers felt that they should be helped to evaluate environmental education materials for themselves. Such evaluations could include factors like cost, suitability for being photocopied, design, from where it could be obtained, appropriateness for particular age groups, readability and use of language. Importantly, information of this type could also be used as guidelines for providers of materials.

A supplementary, but not insignificant, addition to the support materials that have been produced as part of the response to the needs assessment undertaken through the *Environmental Education Information Needs Project* has been the production by the CEE of regular Information Leaflets. These are sponsored by the Esso Petroleum Company and written in collaboration with appropriate member organisations such as the Geographical Association, Association for Science Education, The Tidy Britain Group and The Learning through Landscapes Trust, depending on the topic. They are of low cost to schools and so far include these titles:

- *An introduction to the National Curriculum (England and Wales) and Environmental Education.*
- *Environmental Education within the National Curriculum.*
- *Pollution - Resources for Teachers.*
- *Waste, Litter and Recycling - Resources for Teachers.*
- *Environmental Education through the School Grounds.*
- *Environmental Education through Geography.*
- *Environmental Education through Science.*
  (all available from the CEE)

Broadly, each sets out the rationale in a particular area, often providing a wider educational context with special emphasis on the National Curriculum and the way that teachers are currently expected to work. There then follow short selective bibliographies with brief descriptions of each item and there are further lists of resources, all of which are known to the authors and chosen as the most suitable in the field.

There remain several unresolved problems associated with producing lists of extensively evaluated resources. These partly relate to the logistics of assembling a range of resources representative of any particular field and then ensuring that suitable evaluators are available. It is partly a problem of regular up-dating such lists and partly of ensuring that it is representative of all providers of the type of resource under consideration. This is important because many of these resources are provided by voluntary organisations dependent on their publications to maintain their own memberships.

INSET FOR ENVIRONMENTAL EDUCATION

When the National Curriculum was in the early stages of development it soon became apparent that there was an expectation that areas of education, like environmental education, would be covered mainly through the traditional curriculum subjects. To achieve this the teachers of each subject would identify their particular contribution to the whole. The problems that this raised were soon evident. These related mainly to:

(a) Coherence across the curriculum, in which teachers from different subject areas contributed to the provision of an overall experience for their pupils without there being undue repetition and the way that topics were taught in one subject should support others.

(b) Progression, in which the sequencing of environmental education activities across all of the subjects over the course of the year and from one year to the next should provide experiences which satisfactorily meet the changing intellectual needs of the students.

Both of these considerations raise issues for the regular class teachers as well as for those who have the task of co-ordinating environmental education in the school.

The CEE recognised the various needs and set about providing approaches to in-service training based on the ideas of environmental education as set out in *Curriculum Guidance No. 7* (National Curriculum Council, 1990). Under the sponsorship of National Power and PowerGen they initiated a project developing INSET materials. These conformed to a particular model and had a specific set of aims. They are summarised (after Gayford, 1994) as follows:

- Since the National Curriculum was seen as a priority by most teachers, the materials were designed to help with the implementation of environmental education in this context. Consequently there was an emphasis on the contribution of environmental education to individual subject areas and *vice versa.*
- The materials and approaches would be sufficiently flexible to meet the needs of teachers from different backgrounds and expertise.
- They would be constructed to satisfy a variety of patterns of in-service training to include school-based activities which may form part of either a programme of statutory training days, or an organised programme from an external agency such as a local education authority, higher education institution or NGO.
- They would place environmental education in the context of the whole curriculum. In this way providing opportunities for teachers who are not environmental specialists to contribute to the planning of a whole and coherent environmental education experience for their students.
- Interdisciplinary co-operation would be promoted within schools and assist with progression. Teachers would be encouraged to learn from each other through dialogue between those with expertise in different subject areas, even from different schools and if possible from different phases of education. The concept of peer learning between teachers was considered an important part of the project.
- They would enable and encourage teachers to explore a variety of approaches to, and methodologies for, the delivery of environmental

education, including the context and particular contribution of their own specialism.

- Materials would be provided which will suit the requirements of in-service leaders who may not themselves be experts in environmental education.

Ways in which the development of the materials could best enable these aims to be achieved was thought to be through a modular structure, thus giving them considerable flexibility. The materials were devised to suit the variety of in-service training patterns set up in the different education authorities across the country. The model adopted consisted of an Introductory Module, which was designed to address aspects of environmental education which were likely to be common to teachers in the early stages of introducing an environmental dimension to their teaching. There then followed further modules, each dealing with the needs of teachers who wish to go further in integrating environmental education into their specific subject areas. To date six modules have been published:

(1) *INSET for Environmental Education 5-16: Introductory Activities,* 1992.

(2) *INSET 5-16: Environmental Education for Science,* 1992.

(3) *INSET 5-16: Environmental Education for English,* 1994.

(4) *INSET 5-16: Environmental Education for Geography,* 1994.

(5) *INSET 5-16: Environmental Education for the Arts,* 1995.

(6) *Environmental Education for Initial Teacher Education,* 1995.

A further one, relating Environmental Education to Technology Education, is due to be published in late 1995.

Modules consist of between six and eight activities, each activity usually taking between an hour and ninety minutes to complete. Thus it is possible to run several activities in a single day, sometimes using materials from different modules. In this way leaders of training sessions could select those activities which meet the particular needs of the group. Themes recur across the modules, such as the relationship between environmental education and traditional subject areas, ways of encouraging teachers who represent different curriculum areas to collaborate, the particular contribution that each subject area can make to environmental education, progression and evaluation of resources. These themes are developed in different ways in each module to suit the specialist needs they address.

Approaches within the individual activities vary one to another; some deliberately use methodologies which can be easily adapted for class use with students, giving teachers the experience and confidence to use the approach at a later date in their schools (see Figure 6.1). Other activities are devised to encourage teachers to focus on other aspects of their professional development or on areas of school management or policy formation (see Figure 6.2). Within *INSET 5-16 Environmental Education for Geography,* it is recognised that in many schools, particularly in secondary schools, the senior geography teacher is likely to be the overall co-ordinator for environmental education, and activities have been devised in this module for this particular purpose.

Emphasis has been upon producing materials which are both interactive and encourage teachers to reflect upon their own experience and practice. Accompanying each module there is a tutor's booklet which outlines the general thinking behind each module, with guidance on how to organise the sessions.

The trialling of the materials was a particular feature of their development. This was done in as may contexts as possible in partnership with several local authorities. Formative evaluation following each trial session took the form of discussion with participants as well as the group leaders.

Following on from the recommendation in the findings of the *Environmental Education Information Needs Project* that environmental education should be an integral part of pre-service training courses for teachers, one of the most recent modules, *Environmental Education and Initial Teacher Education,* has been devoted to that area. The fact that initial teacher education in England, Wales and Northern Ireland has been subject to radical changes over the last few years was seen as an additional reason for providing these support materials.

One of the most significant changes that is likely to have an impact on the experiences of teachers 'beginning' environmental education is the amount of time that student teachers are required to spend in schools. This has been significantly increased in the new requirements and is particularly marked in the postgraduate courses which provide training for nearly all secondary teachers. Some of the student's time will be spent on school-based activities which are separate from normal class teaching. Another change is that higher education institutions have entered into partnerships with local schools in a way which provides more opportunities for integration between the experiences of students in their colleges and their schools (DFE, 1991 and 1992; Haggarty, 1995).

As part of many of the new partnerships that have been set up there are school-based mentors whose role is to provide activities for groups of students, often from a range of different specialist disciplines. In many cases this is supported by cross-discipline seminar groups in the college. There is also the subject teaching methodology component of the course which goes on, again both in the schools and colleges. These aspects of the course all

'Bridges or Beasts'
A problem-solving Activity

Summary

This is a practical problem-solving exercise in which participants first construct an animal or bridge from the materials supplied. They are then asked to consider either the adaptation of their animals to changing environmental conditions, or the environmental impact of structures such as bridges on the environment.

Outcomes

By the end of the activity participants will have:

a) used problem solving in the context of environmental education and science education;

b) developed a practical approach to explore some of the implications of environmental change.

Guide Time 70 or 80 minutes

Figure 6.1 Some activities use methodologies that can be easily adapted for class use (example from the Science Module)

Progress Towards a School Policy

Summary

Working in groups, participants are asked to provide replies to a number of hypothetical letters and memoranda that have arrived in the head teacher's in-tray. The letters focus attention on some of the issues which should be considered when developing a school policy for environmental education.

Outcomes

By the end of the activity participants will have:

a) identified important considerations when devising a whole-school policy for environmental education;

b) considered possible strategies for establishing such a policy.

Guide Time 60-75 minutes

Figure 6.2 Some activities help teachers to consider areas of school management and policy (example from the Introductory Module)

provide potential opportunities for pre-service students to experience and practice many aspects of environmental education. An important part of the planning behind this module is that it should reflect and support these new approaches to initial teacher education. Some of the activities have been devised to provide alternative approaches to the same topics. For instance one approach may be to support tutors working with a group of students in the college environment, another may be for a tutor or school-based mentor with a group of students in school and a third may be for a single student or a pair of students working independently.

There have been other recent initiatives, not least is the collaborative project, *Planning and Evaluation of Environmental Education,* shared between the CEE, WWF UK and The University of Reading (Dorion 1993a and b). This set out to address the professional development of teachers in environmental education in primary and secondary schools. The objectives (After Dorion, 1993a and b), were to:

(1) clarify the relationship between environmental education and the National Curriculum subjects;

(2) identify approaches for reviewing and evaluating current provision and planning for environmental education in schools;

(3) identify ideas and approaches for the assessment of students' achievements and experiences in environmental education.

Handbooks were produced, one each for primary and secondary schools. These provide a framework to enable teachers to come to an understanding of the aims and objectives of environmental education in the context of the National Curriculum and their school. It goes on to suggest an approach which will review and evaluate current practice and develop an effective action plan. It finally helps teachers to explore ideas for the assessment of students' achievements and experiences. A process approach is used in which the teachers, individually or in groups within the same school, are invited to review their own practice and use this as a basis for further planning and evaluation of the environmental education in their school.

The material works on the premise that in all schools there is some environmental education going on and the first stage in developing this area of the curriculum is to help teachers recognise what is already happening; then to reflect on this and work towards ways of increasing and improving their practice. Three dimensions are identified to help focus attention on practice. These are summarised under the headings of *what, how* and *when.* Briefly these are respectively, the content, the methodologies used and the context in which it takes place. Rather than simply providing an instructional handbook the materials are constructed as a series of suggestions for action and prompts for reflection on practice, with examples based on practice from contributing schools.

CONCLUSION

The field of in-service education for teachers in the UK, has seen many major changes over the years and teachers have also been expected to respond to a great deal of change in their day to day professional work. It is perhaps easy to assume that with all of the change that is taking place in education, and society in general, that environmental education has a rather low priority in our schools. However, many teachers consider that it is an essential part of their wider role to prepare their students to be responsible citizens. While, understandably, they are not anxious to see further substantial changes or additions to the formal curriculum they do seek more assistance in incorporating environmental education into their day to day teaching and non-formal school activities. They see training as one of the most significant ways of helping them to achieve this aim. The challenge for those seeking to support environmental education in schools remains one of targeting help where it is most useful. To achieve this it is important that a dialogue is maintained between both the providers and users of these services.

ACKNOWLEDGEMENTS

In compiling this chapter I wish to acknowledge the valuable assistance of the staff of the Council for Environmental Education, particularly Christine Midgley (Principle Information Officer) and Maggie Smith (Education Officer, Schools and Colleges).

REFERENCES

Council for Environmental Education 1992. *Environmental Education Information Needs* Reading: CEE.

Department for Education 1992. *Grants for Education Support and Training 1993-4* London: DFE.

Department for Education 1992. *Initial Teacher Training (Secondary Phase)* Circular No. 9/92 London: HMSO.

Department for Education and Science 1991. *School-based initial teacher training in England and Wales: a report by HM Inspectorate* London: HMSO.

Department of the Environment 1991. *This Common Inheritance: Britain's Environmental Strategy* CM 1200 London: HMSO

Department of the Environment 1994. *Sustainable Development: The UK Strategy* CM 2426 London: HMSO.

Department of the Environment 1995. *British Government Panel on Sustainable Development: First Report* (Convenor: Sir Crispin Tickell) London: DoE.

Dorion, C. 1993a. *Planning and Evaluation of Environmental Education for Primary Education* Surrey: WWF UK.

Dorion, C. 1993b. *Planning and Evaluation of Environmental Education for Secondary Education* Surrey: WWF UK.

Gayford, C. G. 1991. Environmental Education: a question of emphasis in the School Curriculum. *Cambridge Journal of Education 21* (1), pp 73-79.

Gayford, C. G. 1994. *Environmental education 5-16: in-service training (INSET) for teachers. Journal of Biological Education 28* (4), pp 284-90.

Haggarty, L. 1995. The Complexities of Effective Mentoring in Initial Teacher Education. *Mentoring and Tutoring 2* (3), pp 32-41.

Hopkinson, G. 1991. The changing arrangements for in-service training. *Annual Review of Environmental Education 4* pp 15-16.

Matthews, G. & Stephens, D. 1992. *Environmental information needs. A report of a survey of environmental information needs in schools and youth work* Reading: Council for Environmental Education

National Curriculum Council 1990. *Curriculum Guidance No. 7: Environmental Education* York: National Curriculum Council

Office for Standards in Education 1993. *Handbook for the Inspection of Schools, 2nd Edition* London: HMSO

Saunders, L., Hewitt, D. & MacDonald, A. 1995. *Education for Life: the cross-curricular themes in primary and secondary schools* Slough: NFER.

Schools Curriculum and Assessment Authority 1994. *The National Curriculum and its Assessment: Final Report* London: SCAA

Sterling, S. 1992. Mapping Environmental Education. In: Filho, W. L. & Palmer, J. A. (Eds.) *Key Issues in Environmental Education* England: Horton.

Smyth, J. C. 1995. Environment and Education: a view of a changing scene. *Environmental Education Research 1* (1), pp 3-20

United Nations 1992. *U.N. Conference on Environment and Development: Agenda 21* Geneva: U.N.

CHAPTER 7

TEACHER TRAINING FOR ENVIRONMENTAL EDUCATION IN GERMANY: TOWARDS A MORE HOLISTIC APPROACH. NATIONAL TRENDS AND EXPERIENCES AT HAMBURG

Klaus Schleicher

ABSTRACT

Since policy regards environmental education as of primary importance (WBGU, 1995) and since environmental education depends more on teacher qualification than on school curricula, one would expect that adequate pre- and in-service training has been well developed. In Germany, however, teacher training not only followed scientific and public debates but also responded to experience at the school level. This delay is due, in part, to the disciplinary structures of the universities (Lübbe, 1987, p 33; Kocka, 1987, p 10) and their slow 'greening'. Furthermore, it results from the fact that the natural and not the social sciences led environmental discussions, and above all, limited evidence exists about predictors that influence teaching competencies and results (Rheinberg, 1986, p 292; OECD, 1990, pp 72 ff). On the whole, information on ecological problems and about technical protection means it is much easier to gain and explain than to change attitudes and behaviour (de Haan, 1994). Generally environmental education has to cope with two complex systems at the same time: on the one hand environmental complexity and on the other educational complexity (Schleicher, 1995a).

In the face of such difficulties, environmental courses of study, respective syllabi or regulations for environmental education should not be tied down. Without giving reason for the conceptional and structural delayed teacher training this appears too easy-going an approach. Accordingly, this chapter focuses first on developments towards a more holistic comprehension of environmental education (first section), then on staff and student awareness of training problems (second section), before empirical evidence about training opportunities is presented (third section) and initiatives are discussed to improve environmental training (fourth section). The final section summarises necessary and possible future developments.

INTRODUCTION

The importance of teacher training for environmental education has been stated at the *international level* since UNESCO's Tbilisi and Moscow conferences in 1977 and 1987 (UNESCO-UNEP, 1985, pp 2 ff; Baines, 1989, pp 40 ff), and *in Germany* since the federal government demanded an integration of more interdisciplinary training and teaching concepts (Bundesregierung, 1972). Meanwhile, the view spread in Germany that information about actual problems is of less importance than transferable ecological insights and competencies (Schleicher, 1995a, pp 10 ff). Accordingly teachers should be offered a synopsis of natural and human ecology, because an understanding of and action on the 'principle of retinity or sustainability' demand value decisions. A synopsis is all the more necessary due to the interaction between the different levels of system-variables (biological, economic, social, *etc.*) (Eulefeld, 1992a, pp 33 ff; Sachverständigenrat, 1994), and teachers need to be qualified with regard to specific target groups, local circumstances and regional curricula.

Teacher training responded to social and school needs respectively after great delay. However, rapid change is vital since environmental knowledge and problems change faster than teaching institutions, school examinations - and, above all, human behaviour. Clearly, *more flexible training structures* and a concentration on *key-qualifications* are therefore required. However, rather quick, simple and always applicable solutions are not feasible because teacher training is faced with complex, psychological, social, environmental and political challenges. Considering this, training has to develop new, transdisciplinary, inter-methodological and multifaceted sequences of training concepts as well as structures. To date, however, there exists limited empirical evidence with regard to environmental *socialisation, attitudes and perceptions* of learners and teachers (Schleicher, 1995b), or how to cope simultaneously with competing ecological and socio-economic demands (Krol & Karpe, 1995).

As it stands, improvements of teacher training concepts in Germany may start by interrelating analyses of environmental education at the school level (Eulefeld, 1992b) with proposals for professional training of environmental advisory services (Graeßner *et al.*, 1987; Baur *et al.*, 1989) and with practical initiatives at the in-service level (*e.g.* Lippke, 1990; Marek, 1993). In addition some hints may be gathered from school based programmes encouraged by OECD (Posch, 1994), from projects supported by the joint Federal and Regional Commission in Germany (BLK, 1991; 1992), from initiatives such as 'Science across Europe' (Association for Science Education, 1991) or from reports on the 'greening' of higher education (Ali Khan, 1995). To date, however, apart from relatively vague references to teacher training (*e.g.* Jaritz, 1993), hardly any complex training concepts have been developed and evaluated (Schleicher & Fricke, 1994). Even the training prerequisites have hardly been clarified (Kahlert, 1991; Krol, 1991).

Thus, it is easy to understand why teacher training follows but does not lead developments in environmental education and educational policy. It will be evident from the following sections on change in environmental education and on empirical evidence, what courses and concepts have been and are still available in teacher training.

DEVELOPMENT TOWARDS A MORE HOLISTIC UNDERSTANDING

Although an increasing number of people and institutions took interest in environmental conservation before the 1950s (Schoenichen, 1954; Zwanzig, 1962), the ecological approach carried little reputation (Van Klaauw, 1965, pp 145 ff). This was due to its restriction to the fields of biology and to the conceptual traditions of university disciplines which hindered interdisciplinary studies. Since the mid-1950s, however, school subjects (primarily biology and geography) paid growing attention to nature protection and landscape conservation. The first university Chair for nature protection was also established and the co-ordinating Conference of the Cultural Ministers of the 'Länder' (Kultusministerkonferenz, KMK) set out guidelines to acquaint students with nature protection and careful landscape treatment (KMK, 1953). However, these initiatives were narrow in scope and hardly incorporated into teacher training.

It was not until the start of the 1970s that a great swing toward environmental concern took place in politics, public opinion, research and education (Kepplinger, 1989). Pollution became a public concern within a year (1969-70), due to large media campaigns and pressure from local action groups. The federal government regarded environmental consciousness, education and even teacher training as an essential element of its first environmental policy programme. Soon *air-*, *water-* and *soil pollution* became almost 'teaching-classics' in biology and geography at university and at school level. In the mid-1970s, when educational policy began to discuss social and economic causes for environmental damage, more comprehensive didactic concepts gained attention (including medical, juridical and political dimensions) and a political dimension in environmental education was demanded for teacher training (Bundeszentrale, 1974, pp 9, 52 ff).

However, it was not until 1980, when a considerable spectrum of problem-approaches, case- and field studies, as well as integrative learning concepts were developed in schools, that the Standing Conference of the Cultural Ministers of the 'Länder' recommended environmental education as a subject-bridging *teaching principle.* Moreover, students at all levels should be acquainted with the interdependence of ecological, economic and social factors, and teachers should be trained with those didactic and interdisciplinary goals in mind (KMK, 1980; 1982; 1987). In this decade public concern and engagement, supported by NGOs, strongly favoured pollution control; some school administrations began to allow *extra-time* or *extra-curricula activities* for environmental education (*e.g.* Baden-Württemberg, 1990); a problem-, situational and conflict oriented teaching spread in schools (Bolscho *et al.*, 1994, pp 22 ff); and slowly transdisciplinary studies became more common at university and teacher training level (Döbler, 1994, pp 27 ff).

In the 1990s, however, the silent revolution in Eastern Europe, Maastricht conflicts and the growing rate of immigration and unemployment withdrew attention from environmental problems (Piel, 1995, pp 372 f). Nevertheless, scientific bodies such as the German Science Council (Wissenschaftsrat) point out that research on interrelations of social and economic subsystems with biotic and abiotic systems are urgent and that greater engagement of the social and economic sciences is overdue, since *"values, attitudes and behaviour of individuals, groups and societies determine the relations between man and society"* (Wissenschaftsrat, 1994a, pp 3, 8, 21 f; Sachverständigenrat, 1987, p 53).

On the whole, environmental education has largely followed political and public discussions and time-specific priorities. Nevertheless it has made considerable progress over the last twenty years. In fact the focus of environmental education changed from nature conservation, in its narrow sense, to pollution control, before broader ecological concepts (*i.a.* a Lernortdidaktik) were envisaged (Schleicher, 1992, pp 29 ff; Salzmann, 1989). Thus, learning concepts, strategies and processes have become increasingly multi- or interdisciplinary, and are increasingly based on experience and addressed to life related issues. Lately this has been accompanied by a 'greening' of learning institutions.

STAFF AND STUDENT AWARENESS OF TRAINING PROBLEMS

In spite of the stated aims of environmental education and the development of environmental awareness in the last decades, universities have not yet developed adequate ecological teacher training concepts that can be based on empirical evidence (Lippke, 1990; Schleicher, 1994, 15 ff). If scientific and public awareness and contexts are to be considered, if the environmental experience and interest of students are to be inculcated, and educational competence to lead to action, then there is a definite need for more information about environmental socialisation, learning prerequisites, teaching inputs, institutional circumstances and, last but not least, about behavioural effects of different concepts and methods. However, it was not until 1989 that a working group of the IPN (Institut für die Pädagogik der Naturwissenschaften) at Kiel started to publish various empirical studies (Eulefeld, 1992c; 1993).

Meanwhile, a number of studies analysed the extent to which teachers feel trained and qualified for environmental education and which obstacles they face in their daily routine. The following examples give a clue as to what was expected either from university or in-service training. According to teachers at the city of Münster, almost 57 % (from different school types) feel handicapped by the principle of subject teaching and/or missing 'minority time'. These deficits were below average with primary school teachers and beyond proportion with teachers of biology, although the subjects of biology and geography were regarded as particularly relevant to environmental education (Hellberg-Rhode, 1992, pp 83 ff, 104). Another study in North Rhine-Westphalia showed that, on the whole, no more than a quarter of interviewed teachers felt competent to teach environmental education, with 50 % citing inadequate training as the cause (Krol, 1991, p 148).

Although interdisciplinary approaches gained popularity at the school level between 1985 and 1991[1] (Bolscho *et al.*, 1994, pp 22 ff), interdisciplinary teacher training was limited till the 1990s (Klenk, 1987, pp 146 ff). Accordingly, it was only reasonable that teachers demanded more interdisciplinary training and that students pleaded for special university courses on environmental education (Hönigsberger, 1991, p 44).

On the whole, a national analysis states that the context for environmental education at the school level improved considerably (Elger *et al.*, 1994, p 20).

1 The *teaching structure* became more relevant in the way that interdisciplinary approaches gained popularity (with an increase from 16 % in 1985 to 21 % in 1991) and extended study-periods became more common. The percentage of those lasting 3 to 5 hours increased from 27 % to 32 %. Those with at least 6 hours more than doubled (from 3 % to 8 %) (Bolscho *et al.*, 1994, pp 22 ff).

Nevertheless, environmental education still seems to focus on changes in personal life-styles and has not become an overall 'educational principle'. Furthermore, interdisciplinary approaches are not yet widely spread. The often used 'hands-on' approach rarely leads beyond some actionism and, finally, social, political as well as economic dimensions are not very well integrated in environmental studies. It is likely that many teachers feel that *active learning, interdisciplinary studies, interrelation to out-of-school reality or concern for sustainable development* expects too much from them. Often they, as well as the trained students, express more environmental interest than environmental consciousness and knowledge (*ibid.* p 21; see Figure 7.1, *ibid.* p 24).

Figure 7.1 Environmental information, awareness and interest/engagement

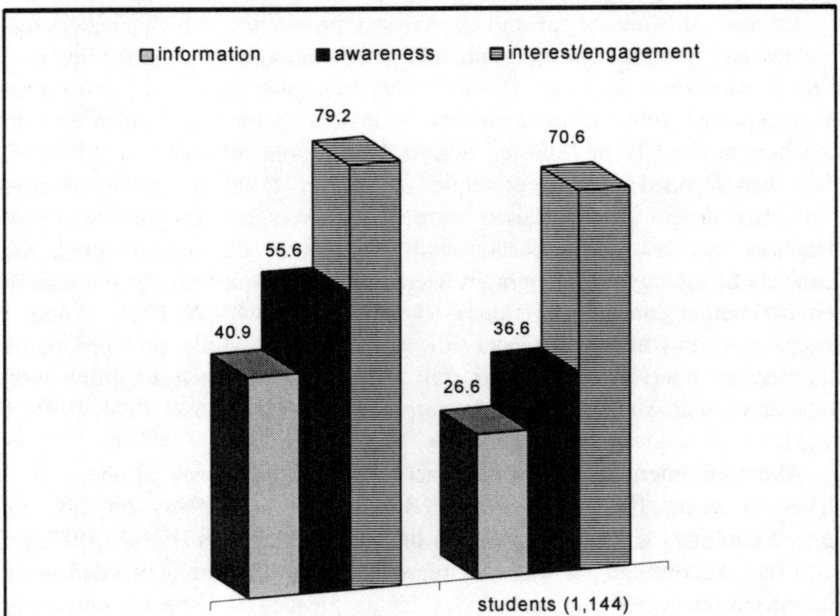

Beyond this, both groups stated that mass-media offered the most important information while university teaching or in-serve training ranked almost last. Accordingly they regarded media, economy and environmental associations

as much more important for environmental problem solving than school education or universities (*ibid.* p 26).

Thus, because the majority of teachers received their initial training almost 20 years ago, the quality and practice of today's environmental teaching in schools seem to result more from private initiative and public or curricular demand rather than from conceptional changes in teacher training.

EMPIRICAL EVIDENCE ON TRAINING OPPORTUNITIES

In Germany, teacher training is based on three components: educational theory, subject-related didactics and at least one traditional university discipline. Although all components may include environmental dimensions, the final qualifications of teachers differ according to the mix of components chosen and disciplines studied. Such study options certainly depend on the extent to which environmental dimensions became part of university courses.

Additional information about teacher training and in-service training is available from a nationwide analysis of university courses (1991 / 92), staff surveys (899 members) and questionnaires administered to students of two teacher training colleges (1,144), and an additional analysis of training programmes of in-service institutions. The study (Elger *et al.*, 1994) shows that 56 universities offer at least some environmental courses in the field of teacher training (*ibid.* p 42). However, about two thirds of the courses are offered in biology (43 %), geography (16 %) and chemistry (8 %), while education courses fall short at less than 10 % (7 %) (*ibid.* p 49). If one differentiates further with regard to the subject, didactic and educational focus of the courses, it becomes obvious that university courses are much more related to disciplines while colleges of education are more concerned with environmental didactics (Figure 7.2, *ibid.* p 53).

The emphasis on particular study components (educational theory, didactics, disciplines), together with the uneven emphasis on interdisciplinarity in social and natural science has led to three important consequences in teacher training:

(1) At university level, no complete study courses on 'environmental education' are available or examined.

(2) Many ecological courses are offered, although they are not a sufficient substitute to those in environmental education.

(3) Finally, since environmental education is regarded as a teaching principle rather than a subject at school level, adequate training is obviously missing (Entrich *et al.*, 1995, pp 12 ff).

Therefore considerable change in pre-service teacher training is necessary if these discrepancies in scope, methodology and reference are no longer to be a burden to teachers

Figure 7.2 Main focus of study courses for training in environmental education

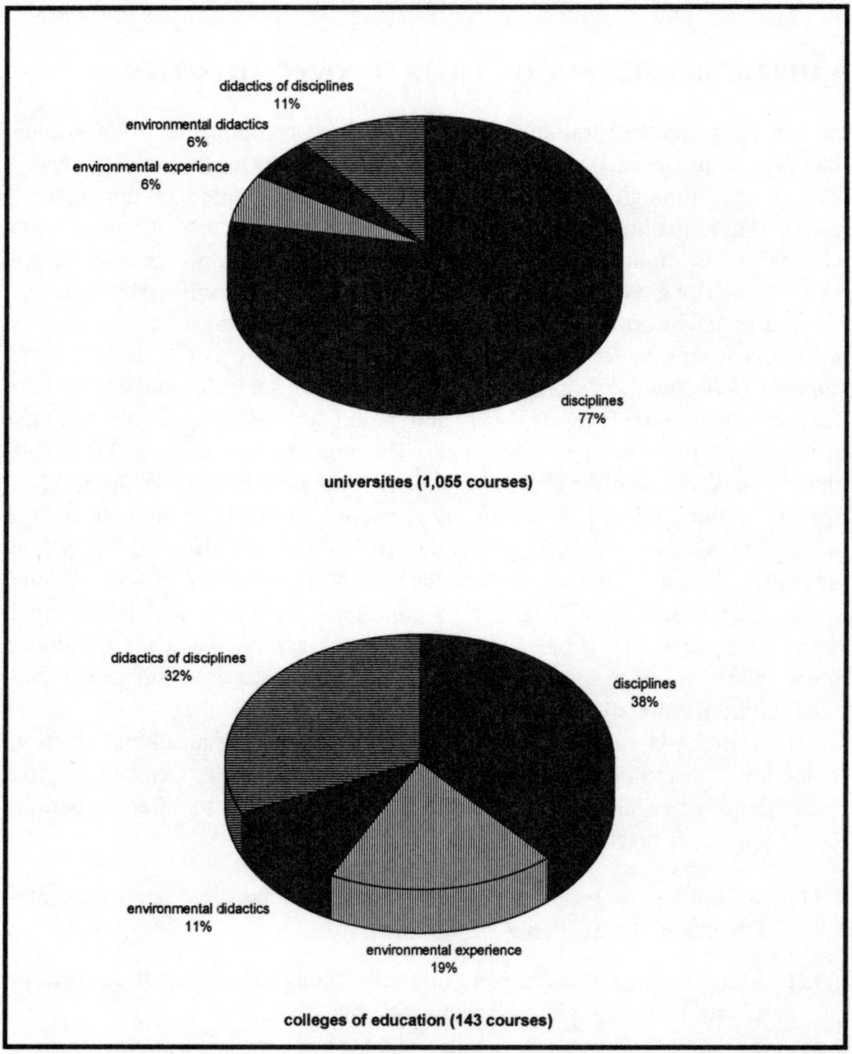

These training deficits are not compensated by in-service training, which is organised by special institutions of the 'Länder'. Though some of the

'Länder' have chosen environmental education as a main topic for in-service training (as Baden-Württemberg, Bavaria and Berlin) most of the in-service training is still related to school subjects (88 %), i.e. comparatively little attention was paid to pedagogical and methodological aspects (12 %) (Elger, *et al.* 1994, pp 66, 71). More complex approaches have only been envisaged in two experimental in-service training projects (Berlin, Saarland) which tried to organise regional networks based on transdisciplinary practical approaches (*ibid.* p 76). In addition an interdisciplinary in-service programme has been offered at Oldenburg since 1984 and a preliminary two-semester course has been tested at Hamburg from 1991 to 1995. Finally, follow-up studies are offered in Ludwigsburg at the college of education since 1992 although the foundational concept is still experimental (Eulefeld, 1992a; Bolscho *et al.*, 1994, pp 43 ff).

INITIATIVES TO IMPROVE ENVIRONMENTAL TRAINING

In the face of recent developments (growing public concern for environmental protection since the 1970s; increasing attention paid to holistic concepts in environmental education since the 1980s; and a gradual 'greening' of educational institutions in the 1990s) it became obvious that drastic changes in teacher training were required to meet the demands of educational policy and curricula guidelines as well as the claims made by teachers and students. Unrelated environmental classes offered by various disciplines and subject-didactics were increasingly regarded as un-ecological and ineffective. Obviously teacher training somehow needed a 'holistic approach' considering natural and social dimensions of environmental development as well as the relevance of learning inputs from the biographic environment.

Based on these demands two major model-projects were launched on 'environmental education and consultation' at Hamburg University. For the first time the joint 'federal and regional commission for innovation in education' (Bund-Länder-Kommission) financed a programme to improve concepts of teacher training in environmental education at university level (after many earlier projects on school teaching).

The goal of the first of these projects *(1989-1992)* was: to analyse available training initiatives and ideas; to interrelate natural and human ecology with regard to 'retinity'; to specify environmental situations in urban areas suitable for training; to develop a concept for interdisciplinary environmental didactics; and to start experimental 'contact studies' for teachers (Hoebel-Mävers *et al.*, 1991; Schleicher, 1992). Certainly not all hopes held out (Döbler, 1995) but on the whole, considerable interest and many initiatives amongst staff and students were raised. Finally, the education faculty resisted the implementation of a study course in environmental education although the German Science Council welcomed 'the rare environmental impetus' at Hamburg compared to other German universities. Beyond this acknowledgement the Council called that more concern should be given to dynamic interactions between man and environment, that more attention should be paid to the psychology of environmental learning and that more empirical evidence should be provided (Wissenschaftsrat, 1994b).

Based on these experiences and criticisms, the second project *(1994-1997)* is conceptualised as an inter-university and cross-disciplinary training scheme between the 'Länder' of Hamburg and Thuringia. It should encourage and improve environmental awareness and training in one of the 'Neue Länder' and feed back its experiences with recently set up training schemes to Hamburg. Both universities seem to gain from the co-operation, since a conceptual and structural comparison of their training intentions points not only towards the political, socio-economic and natural diversity of their regions, but also towards the different university structures, study courses, staff qualifications and student interest. The project will show to what extent a holistic structure, content and learning process in teacher training may be developed.

On the one hand, the transfer-concept demands a rather flexible and dynamic structure, with training segments that can be combined in various degrees, and with small modules that can be used in different study contexts. This flexibility would serve small student groups aiming at different qualifications (teacher certificates, adult education diplomas or magister examinations). Beyond this, it allows easier adaptation to new research findings or current environmental situations without changing whole study courses.

On the other hand, it is essential to initiate an effective learning process so that scientific, teaching and learning prerequisites are clarified, for instance with regard to the importance of environmental socialisation, in view of a more holistic didactic approach to environmental learning, or in the face of students' knowledge and engagement. Accordingly, the following issues were analysed in Hamburg, in various papers:

- to what extent concepts for environmental education have been incorporated into *educational theory* (Schleicher, 1995a);
- how far *family socialisation* effects environmental attitudes and learning (Schleicher, 1995b);
- how holistic *learning environments* may be chosen by specific criteria (Schleicher, 1992);
- which *environmental theories* allow what kind of environmental decision making (Döbler, 1992);

- to what extent *study structures* in various areas permit more or less interdisciplinarity (Döbler, 1994);
- in what way *environmental awareness, comprehension and competence* differ amongst students in the Eastern and Western region (Möller & Fricke, 1995; Fricke & Möller, 1995);
- which discrepancies exist between *training goals of staff and students* (Schleicher & Fricke, 1995). An example is given in Figure 7.3 (Schleicher, 1994, p 131).

Figure 7.3 'Abilities' to be promoted in environmental education courses

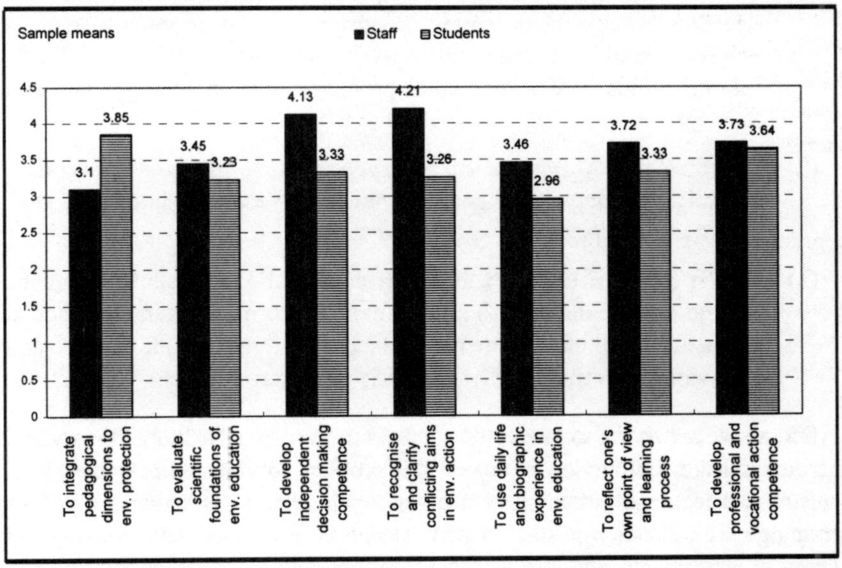

Training intentions of the staff were compared to the students' recognition of these with the threefold aim: first, to improve the match between seminar content, didactic settings and social contexts; secondly, to increase the students' self-awareness with regard to their study needs and professional goals; and thirdly, to stimulate research in university didactics (Schleicher & Fricke, 1994; 1995).

Finally, an analysis was undertaken to assess the relevance of teaching / learning theories, contents and methods and environmental, social and institutional contexts to the development of environmental competencies. Until now, interdisciplinary lectures were offered to provide insight into the importance of disciplinary paradigms for environmental education (*e.g.* legal, economic, technical or of different social sciences) (Möller, 1995). Furthermore, an evaluation was made of the degree to which seminars can offer methodological expertise in dealing with school and out-of-school environments, with environmental risks and conflicts, or with different paradigms and environmental ethics (Döbler, 1992). These experiences were followed up by a study of those *environmental competencies* that are important to environmental maturity. That is, what process oriented qualifications are regarded as important in teacher training to date.

At Hamburg University basic ideas are tried out under the following premises:

(1) To what extent *overviews on disciplinary studies* facilitates practical decision making with regard to environmental problems (*e.g.* in environmental planning), respectively how useful are disciplinary interpretations of environmental problems (*e.g.* in biology, law and history).

(2) In which way *sequential study structures* have to allow conceptual, didactic and competence goals, or have to give way to interest of staff and to continuity of study groups of students.

(3) Which job profiles demand which combinations of study elements. On the whole, students should be exposed to both natural and social sciences, with environmental didactics and also with values and behaviour relevant incentives to sustainable development.

All in all we are trying to interrelate subject knowledge, educational theory and student engagement in view of responsible problem solving competencies, with regard to vocational options and in order to stimulate a 'greening' of teacher training. The question is still, to what extent an inter-university exchange of study programmes is possible, whether a progressing sequence of learning can be combined with the use of transferable modules and how staff qualifications and teaching interest can be brought in line with student interest and, finally, with professional qualifications.

OUTLOOK TOWARDS FUTURE CHALLENGES AND DEVELOPMENTS

Improvements in teacher training for environmental education have been demanded for over two decades, at both the international and national level. Progress, however, has been slow and limited because the natural sciences have traditionally dominated environmental thinking and their teaching has largely been related to a cognitive, top down profile. Furthermore, university structures and curricula are even more reluctant to change towards interdisciplinary studies than are those of schools. Finally, up until the 1990s, ecological paradigms were barely incorporated into educational science theory. Obviously new approaches are needed, because of the complexity of environmental problems and because education for sustainable development demands considerable co-operation at the theoretical and institutional level, for which no experts are available - except those self-elected.

Nevertheless, points of departure for future training are available from some universities in the 'Neue Länder' and from technical universities. They at least have tried to co-ordinate the considerable amount of environmental courses supplied by various disciplines for environmental training and education. On the one hand, some institutions offering a 'Studium generale' related to environmental aspects have set up university centres for environmental sciences, or have approved certified study-segments in environmental education (*e.g.* Halle-Wittenberg: Hartmann, 1995, pp 128, 148 ff). Others wanted to introduce interdisciplinary environmental segments to all disciplinary course studies (*e.g.* Hannover: Westholm, 1995, pp 131 ff). Beyond this, the Technical University Darmstadt conceptualised an experimental, interdisciplinary curriculum in 'environmental education', although only for engineers. Lastly but not least, the Technical University Cottbus established a 'faculty of environmental sciences' with chairs from quite different disciplines.

In the face of such promising efforts, and additional evidence that environmental information has little influence on 'high cost' behaviour (de Haan 1994), teacher training should and could be more effective if the following two conditions are met:

(1) It should concentrate as much on *pedagogical specifics* of environmental education as on disciplinary knowledge.

(2) It should rearrange or *restructure* some parts of its study courses, syllabi and examinations.

In conceptual terms, it seems to be unrealistic to aim at a really *holistic education concept* (v. Hentig, 1987) or to institutionalise compulsory courses without considering student competence or staff interest (Westholm, 1995, pp 182 f). Instead more attention should, and could be paid to a number of integrative dimensions of environmental courses. The major aim in environmental education is definitely not to expand environmental knowledge for its own sake but to acquaint teachers, for instance, with the 'principle of retinity' and with some 'foundations of sustainable development', so that they can prepare their pupils to cope with rapid environmental, societal and scientific changes. This, however, means that teachers need to know how to cope with considerable uncertainty in the face of open, complex and self-dynamic environmental processes (Bruhn, 1993), and in view of the insoluble tension between scientific knowledge about, and democratic responsibility for, environmental development.

In organisational terms, educational departments need therefore to clarify how they want to interrelate disciplinary, educational, didactic and extra-functional competencies within the study courses offered. The range of different training options which are offered throughout Germany is evident from empirical analyses (Elger *et al.*, 1994). On the whole, educational policy favours structural changes in teacher training syllabi and examinations but pays little attention to the 'greening' of higher education or to university didactics. On the one hand it is suggested that course elements on environmental education should become compulsory elements of various training syllabi and examinations (Sachverständigenrat, 1994). On the other hand more freedom of choice is left to students. That is, training institutions and examining boards are required to encourage specific subject combinations and more interdisciplinary studies (Elger *et al.,* 1994, pp 121 ff). In any case, courses related to *methodological* and *didactic problems* of environmental teaching and learning could help to bridge the gap between the 'two cultures' (Elger *et al.,* 1994, p 117). To date, however, only a few universities are really 'greening' in terms of the *University Charter for Sustainable Development* and its ten principles initiated by the Conference of European Rectors in 1993. Beyond this, the importance of change in university didactics for a more ecological approach, balancing teaching intentions as well as student interest and competence, is still hardly recognised (Schleicher, 1994, pp 121 ff).

In summary, one could argue whether the conclusions of an OECD report on environmental education in 1992 still posit a challenge to teacher training in Germany, since it has been stated that:

> *"Environmental education ... calls for a sincere and unequivocal realignment of educational philosophy... Strategic policies for environmental education can only be seen in a sustainable, long-term perspective"* (Schneider, 1993, pp 98 f).

REFERENCES

Ali Khan, S. 1995. *The Environmental Agenda. Taking Responsibility. Promoting Sustainable Practice through Higher Education Curricula* London: Pluto Press.

Association for Science Education 1991. *Science across Europe. Saurer Regen über Europa (Acid rain over Europe)* Herts, UK: Association for Science Education.

Baden - Württemberg, Ministerium für Kultur und Sport 1990, Umwelterziehung ist Daueraufgabe aller Unterichtsfächer, *Pressemitteilung 17.9*

Baines, J. 1989. Umwelterziehung in Europa - Ein Überblick. In: Bundesminister für Bildung und Wissenschaft: Umweltbildung in der EG. *Studien: Bildung - Wissenschaft 79*, pp 39 ff Bonn: BMBW.

Baur, R., Kaltenbach, J. & Versteegen, D. 1989. *Modellprojekt Umweltberatung für Haushalte und Gemeinden.* Basel: Prognos.

BLK (Bund-Länder-Kommission) 1991. *Projektgruppe 'Innovation im Bildungswesen': Modellversuche zur Umweltbildung* Bonn: BI 19/21.

BLK (Bund-Länder-Kommission für Bildungsplanung und Forschungsförderung) 1992. *Informationsschrift über Modellversuche im Bildungswesen* Bonn: BLK.

Bolscho, D., Eulefeld, G. & Seybold, H. 1994. Bildung und Ausbildung im Umweltschutz. In: Buchwald, K. & Engelhardt, W. (Hrsg) *Reihe Umweltschutz - Grundlagen und Praxis, Bd. 17* Bonn: Economica.

Bruhn, J. 1993. Probleme unserer Zeit als Herausforderung für den naturwissenschaftlichen Unterricht. *MNU 4*, pp 195-198.

Bundesregierung 1972. *Das Umweltprogramm* Stuttgart: Bundesregierung.

Bundeszentrale für politische Bildung 1974. Umweltschutz als fachübergreifendes Curriculum. *Schriftenreihe, Heft 99*, Bonn.

de Haan, G. 1994. *Umweltbewußtsein - ein kulturelles Konstrukt*: Forschungsgruppe Umweltbildung. Berlin: Mimeo.

Döbler, M. 1992. Die ethische Dimension der Umwelterziehung. Konkurrierende Ansätze und pädagogische Aufgaben. In: Schleicher, 1992, pp 327-379.

Döbler, M. 1994. Interdisziplinarität in Umweltstudiengängen. In: Schleicher, 1994, pp 277-317.

Döbler, M. 1995. Umweltbildung im Hochschulbereich. Ziele, Erfahrungen und Perspektiven der Hamburger Modellversuche. *Diskussionspapiere des BLK-Modellversuchs, Nr. 2*, Hamburg.

Elger, U., Hönigsberger, H. & Schluchter, W. 1994. *Umwelterziehung in der Lehreraus- und -fortbildung* Berlin: UNESCO-Verbindungsstelle für Umwelterziehung / Umweltbundes-amt.

Entrich, H., Eulefeld, G. & Jaritz, K. (Hrsg.) 1995. *Fallstudien zu Umwelterziehung / Umweltbildung in Forschung, Lehre und Studium* Kiel: IPN.

Eulefeld, G. 1992a. Von der Notwendigkeit der Netzwerke in der Umwelterziehung. In: Eulefeld, G., Guder, K. & Maltz, A. (Hrsg.) *Modell zur Umwelterziehung in der Bundesrepublik Deutschland, Bd. 4* Kiel: DGU.

Eulefeld, G. 1992b. 15 Jahre Umwelterziehung auf dem Weg zum Ziel? *Bildungsforschung und Bildungspraxis 2*, pp 131 ff.

Eulefeld, G. (Hrsg.) 1992c. *Empirische Studien im Bereich Umwelterziehung Voraussetzungen, Zwischenberichte, Ergebnisse* Kiel: IPN.

Eulefeld, G. (Hrsg.) 1993. *Studien zur Umwelterziehung. Ansätze und Ergebnisse empirischer Forschung* Kiel: IPN.

Fricke, A. & Möller, Chr. 1995. Ausbildungsaufgaben in der Umweltpädagogik. Erwartungen von Hamburger Studierenden. *Studienmaterialien zur Umweltbildung im Hamburger BLK-Modellversuch, Nr. 9*, Hamburg.

Graeßner, G., Obladen, H.-P. & Schmithals, F. 1987. *Professionalisierung und Qualifikation von Umweltberatern. Ergebnisse einer empirischen Untersuchung.* Universität Bielefeld (IZHD) 1.

Hartmann, E. 1995. Universitäre Umwelterziehung / Umweltbildung für pädagogische Berufe: Das Beispiel Halle-Wittenberg. In: Entrich, 1995, pp 103 ff.

Hellberg-Rhode, G. 1992 Umwelterziehungspraxis im Sach- und Biologieunterricht. Ergebnisse einer empirischen Studie im Schuljahr 1989 / 90. In: Eulefeld, G. (Hrsg.) *Empirische Studien im Bereich Umwelterziehung - Voraussetzungen, Zwischenberichte, Ergebnisse* Kiel: IPN, pp 83 ff.

Hoebel-Mävers, M., Schleicher, K., Gärtner, H. & Schreier, H. 1991. BLK-Modellversuch 'Umweltbildung, Umwelterziehung, Umweltberatung'. In: Gärtner, H. & Hoebel-Mävers, M. (Hrsg.) *Umwelterziehung - ökologisches Handeln im Ballungsraum* Hamburg: Krämer, pp 194 ff.

Hönigsberger, H. 1991. Fachgespräch über Stand und Probleme der umweltorientierten Lehreraus- und -fortbildung. Tagungsunterlagen. Heidelberg: *cf.* BLK 19 / 91, p 44.

Jaritz, K.-E. 1993. Umwelterziehung und Lehrerausbildung. In: DGU *Nachrichten* 7 / 1993, pp 30 ff.

Kahlert, J. 1991. Alltagstheorien der Lehrer über den Zustand der Natur. In: Eulefeld, G., Bolscho, D. & Seybold, H. (Hrsg.) *Umweltbewußtsein und Umwelterziehung* Kiel: IPN 129, pp 65 ff.

Kepplinger, H. M. 1989. Artificial Horizons: How the press presented and the population received technology in Germany from 1965-1986. Paper presented at the Fourth International Congress of PWPA, London 25-29, London: Mimeo.

Klenk, G. 1987. *Umwelterziehung in den allgemeinbildenden Schulen. Entwicklung, Stand, Probleme - aufgezeigt am Beispiel Bayern* Frankfurt M: Haag & Herchen.

KMK (Sekretariat der Ständigen Konferenz der Kultusminister der Länder) 1953. *Naturschutz und Landschaftspflege sowie Tierschutz. Rundschreiben des Sekretariats vom 30.9.1953 (Nr. III 3740 / 53)* Bonn: KMK

KMK 1980. *Umwelt und Unterricht. Beschluß der KMK vom 17.10.1980* (Erg.-Lfg. 44 vom 7.12.1981), p 669 Bonn: KMK.

KMK 1982. *Umwelterziehung in der Schule* Bonn: KMK.

KMK (Hrsg.) 1987. *Umwelterziehung in der Schule. Beitrag zum Europäischen Umweltjahr 1987* Bonn: KMK.

Kocka, J. (Hrsg.) 1987. *Interdisziplinarität, Praxis - Herausforderung - Ideologie* Frankfurt M: Suhrkamp.

Krol, G.-J. 1991. Begründung eines eigenständigen sozialökonomischen Beitrages zur Umweltbildung und Umwelterziehung - Lehrerbefragung zur Umwelterziehung. In: Eulefeld, G., Bolscho, D. & Seybold, H. (Hrsg.) *Umweltbewußtsein und Umwelterziehung* Kiel: IPN 129, pp 135 ff.

Krol, G.-J. & Karpe, J. 1995. Die Relevanz der umweltökonomischen Dimension für eine europäische Umweltbildung. In: Schleicher, 1995b, *Umweltbewußtsein und Umweltbildung in der Europäischen Union*, pp 47-78Hamburg : Krämer,.

Lippke, W. (Hrsg.) 1990. *Unterricht - Studium - Fortbildung. Band 2: Umwelterziehung - Umweltschutz.* Universität Siegen: Praktikumsbüro.

Lübbe, H. 1987. Helmut Schelsky und die Interdisziplinarität / Zur Philosophie gegenwärtiger Wissenschaftskultur. In: Kocka, 1987, pp 17 ff.

Marek, R. (Hrsg.) 1993. *Praxisnahe Umwelterziehung* Hamburg: Krämer.

Möller, Chr. 1995. Die künftige Umwelt erhalten und gestalten. Eine Dokumentation der Vorlesungsreihe von K. Schleicher im Rahmen des Hamburger Modellversuchs. Teil I. *Diskussionspapiere des BLK-Modellversuchs, Nr. 5*, Hamburg.

Möller, Chr. & Fricke, A. 1995. Ausbildungsaufgaben in der Umweltpädagogik. Erwartungen von Erfurter Studierenden. *Studienmaterialien zur Umweltbildung im Hamburger BLK-Modellversuch, Nr. 8*, Hamburg.

OECD 1990. *The Teacher Today. Tasks, Conditions, Policies* Paris: OECD.

Piel, E. 1995. Risikowahrnehmung und Umweltbewußtsein im Spannungsfeld medienvermittelter Wahrnehmung. In: Schleicher, 1995b,

Umweltbewußtsein und Umweltbildung in der Europäischen Union, pp pp369-395 Hamburg : Krämer,.

Posch, P. 1994. The study 'environment and school initiatives': phase one. In: OECD *Evaluating innovation in environmental education* Paris: OECD, pp 21 ff.

Rheinberg, F. 1986. Erzieherforschung am Beispiel des Lehrers. In: Weidemann, B. *et al.* (Hrsg.) *Pädagogische Psychologie* München Weinheim: Urban & Schwarzenberg, pp 279 ff.

Sachverständigenrat 1987. *Umweltgutachten 1987*, Stuttgart / Mainz: Kohlhammer, pp 48-53, pp 407-435.

Sachverständigenrat 1994. Umweltgutachten 1994 des Rates von Sachverständigen für Umweltfragen. Für eine dauerhaft-umweltgerechte Entwicklung. Unterrichtung durch die Bundesregierung. *Deutscher Bundestag, 12. Wahlperiode, Drucksache 12/6995, 8.3.*

Salzmann, Chr. 1989. Regionales Lernen an Lernstandorten. *Die Grundschule 3*, pp 48 ff.

Schleicher, K. 1992. Theorie und Didaktik von Lernorten. Begründung und Anwendung einer Lernortdidaktik. In: Derselbe (Hrsg.) *Lernorte in der Umwelterziehung* Hamburg: Krämer, pp 29-85.

Schleicher, K. 1994. *Umweltbildung von Lehrern* Hamburg: Krämer.

Schleicher, K. 1995a. Umwelterziehung / Umweltbildung in wissenschaftlichen Einrichtungen. Standort und Beitrag der Erziehungswissenschaften. *Diskussionspapiere des BLK-Modellversuchs, Nr. 1*, Hamburg.

Schleicher, K. 1995b. Umwelterziehung und -bildung in der Familie. *Diskussionspapiere des BLK-Modellversuchs, Nr. 7*, Hamburg.

Schleicher, K. & Fricke A. 1994. Umwelteinstellungen von Studierenden. Eine Längsschnittstudie (1989-1992). In: Schleicher, 1994, pp 63-120.

Schleicher, K. & Fricke A. 1995. Ausbildungsaufgaben in der Umweltpädagogik in Ost- und Westdeutschland. Ein Regionalvergleich in Ergänzung zu Längsschnittstudien von 1989 / 1992. *Diskussionspapiere des BLK-Modellversuchs, Nr. 4*, Hamburg.

Schneider, H. (Ed.) 1993. *Environmental education. An approach to sustainable development*. Paris: OECD.

Schoenichen, W. 1954. *Naturschutz, Heimatschutz. Ihre Begründung durch Ernst Rudorff, Hugo Conwentz und ihre Vorläufer* Stuttgart: Wiss. Verlagsgesellschaft.

UNESCO-UNEP (Division of Science, Technical and Vocational Education) 1985. A comparative survey of the incorporation of environmental education into school curricula. *Environmental Education Series No. 17* Paris: prepared by UNESCO Institute of Education Hamburg, o. J.

v. Hentig, H. 1987. Polyphem oder Argos? Disziplinarität in der nicht-disziplinären Wirklichkeit. In: Kocka, J. (Hrsg.) *Interdisziplinarität, Praxis - Herausforderung - Ideologie* Frankfurt M: Suhrkamp, pp 34 ff.

van Klaauw, C. J. 1965. Ökologische Studien und Kritiken. In: *Sudhoffs Archiv für Gesch. der Medizin und der Naturwiss* (First edition 1936) Wiesbaden, pp 145 ff.

WBGU (Wissenschaftlicher Beirat der Bundesregierung Globale Umweltveränderungen) 1995. *Welt im Wandel: Wege zur Lösung globaler Umweltprobleme. Jahresgutachten 1995 (Kurzfassung)* Bremerhaven: WBGU.

Westholm, H. 1995. *Sozial-ökologische Grundbildung an Hochschulen. Umweltbildung als Instrument der Umweltpolitik* Frankfurt: Peter Lang.

Wissenschaftsrat 1994a. *Stellungnahme zur Umweltforschung in Deutschland. Kurzfassung* Schwerin: Mimeo, 20.5.

Wissenschaftsrat 1994b. *Teil C: Regionalteil Schleswig-Holstein und Hamburg* Schwerin: Mimeo, 20.5.

Zwanzig, G. 1962. *Die Fortentwicklung des Naturschutzrechts in Deutschland nach 1945. Bd. 1* Erlangen: Merkel.

CHAPTER 8

M.E.S.A.G.E.S.: THE PROMOTION OF CONCEPTS AND PRACTICES OF ENVIRONMENTAL EDUCATION IN THE CONTEXT OF IN-SERVICE TEACHER EDUCATION PROGRAMMES IN EUROPEAN SECONDARY SCHOOLS

Kathryn O'Loan

INTRODUCTION

"Sustainable living must be the new pattern for all levels : individuals, communities, nations and the world. To adopt new patterns will require a significant change in attitudes and practices of many people. We will need to ensure that education programmes reflect the importance of an ethic for living sustainably",(IUCN, UNEP, WWF, 1991)

Although environmental education is not a new concept, only since the late 1980s and early 1990s has *education for sustainable living* emerged as a focus for this established educational tradition. Education for sustainable living means encouraging people to recognise the need for future development - both economic and social - to occur with due attention to the needs of the environment, which forms our ultimate resource base. In order to achieve this, education, whether it is termed 'environmental' or 'for sustainable living', must involve learners in local issues and the local aspect of wider issues. To this end it is vital that teachers feel confident to tackle environmental education not merely within their own classrooms, but within their own schools.

This chapter seeks to review the development of a secondary level in-service teacher training package for environmental education for the

European Union - MESAGES. To this end, a general view of the impetus for, some of the issues encountered and aims of the end product, will be discussed.

MESAGES - Module on the Environment, Sustainability and Awareness Geared to European (secondary) Schools - is an initiative aimed at in-service teacher education (INSET) for the European Union. Essentially, it is designed for INSET facilitators, to provide secondary level teachers of all subjects with an opportunity to explore and clarify a personal understanding of the term *environmental education* and to develop and apply this to the schools within which they work.

One of the main features of this project was the establishment of the Euro-Awareness network, which consisted of a number of teachers, teacher trainers and educational practitioners from throughout Europe. Both the network and the MESAGES project are co-ordinated by the European Research and Training Centre on Environmental Education (ERTCEE), at the University of Bradford England. On completion of the project the network will remain active and ERTCEE will continue to assimilate feedback. This initiative was partially funded by DGXI of the European Commission.

THE CONCEPT

The guiding force behind the development of MESAGES is the fact that, although environmental education is becoming quite popular throughout Europe, it is not as widespread as is possible or indeed, needs to be, to enable teachers to make their own decisions regarding application in relation to their own weekly or yearly schemes of work. This is due, in part, to the fact that a significant proportion of European teachers do not have sufficient information on the basic concepts of environmental education.

For some time now, attention has been focused upon the role of teacher training in the progress of both the environmental education of students and the achievement of environmental sustainability. As early as the Belgrade Charter, UNESCO (1976), for example, it was stated that environmental education should be a necessary component of pre- and in-service teacher education programmes.

More recently, the 1988 Resolution by the European Council and the Ministers of Education from the relevant European States, suggested that each Member State should "take appropriate measures to develop teachers' knowledge of environmental matters in the context of their initial and in-service training". This document further suggested that each Member State should:

- give consideration to the basic aims of environmental education when drawing up curricula and organising interdisciplinary courses;
- encourage extracurricular activities by means of which theoretical knowledge of the environment acquired in school can be put into practice;
- undertake specific action to provide teachers and pupils with appropriate teaching materials.

Again in 1992, the Ministers of Education meeting with the Council of Europe, reinforced the need for teacher education and development with regard to environmental education, "special attention should be given to the intensification of initial and in-service training of teachers in this area". However, the European Commission has left decisions regarding the direction, content and delivery of such programmes to the individual Member States.

ENVIRONMENTAL EDUCATION FOR TEACHERS : IDENTIFYING OBJECTIVES

In 1990 UNESCO-UNEP made a valiant effort to provide guidelines for the development of teachers in environmental education in a paper titled 'Environmentally educated teachers: the priorities of priorities'. The scope and range of competencies included, clearly places the UNESCO-UNEP programme within the domain of pre-service education, as is illustrated by the sample of competencies outlined below. In this paper, it is proposed that teacher education programmes for environmental education should provide teachers with, firstly, *foundation competencies in professional education* and secondly, *competencies in environmental education content* (Oulton & Scott, 1995).

The foundation competencies focus upon the relationship between general educational philosophies, aims and criteria and environmental education at teacher education level, for example, teachers should:

- "effectively evaluate the results of environmental education curricula and methods in both cognitive and affective domains;
- utilise current theories of moral reasoning in selecting, developing and implementing environmental education curricula...".

Four levels of competence in environmental education content are suggested: ecological foundations, conceptual awareness, investigation and evaluation, environmental action skills. The full range of criteria may even be impossible to achieve in all but the most dedicated of courses because of the time scale involved.

Nevertheless, there are a number of implications for the development of an in-service teacher education model. To begin with, the term 'competency' may not transfer easily to in-service teacher education. Given that most INSET generally takes no longer than one day, it is probably unreasonable to expect teachers (who for INSET are - learners) to become *competent* for environmental education purposes, in this space of time. For this reason MESAGES employs the term *objectives* to mean the activities or focus of activities that participants will undertake during each section.

A number of criteria relating to both the foundation and content competencies within the UNESCO-UNEP programme can be translated into objectives for INSET. Those chosen depend of course upon the view of environmental education held by the INSET designers. Those relating to competence in *investigation and evaluation and environmental action skills* resemble most closely the aims of and objectives contained within, MESAGES (See Table 8.1 for illustration).

MESAGES differs however, from the UNESCO-UNEP programme in the interpretation of these competencies/objectives. While the development of environmental education in the UNESCO-UNEP programme hinges upon knowledge of the ecological and the environmental, in MESAGES it is the professional experience, skill and will of the individual teacher trainer and the individual teacher which serves as the drive for development. This is not to say that one is more important than the other and is perhaps as much a reflection of the relative time scales proposed in each, as of any other factor.

There are a range of other issues which have significant implications for the development of an in-service teacher education model for the European Union. Most teachers, for instance, are not committed environmentalists. Most are however, interested in the welfare, education and holistic development of their students. In this respect teachers will, it must be assumed, seek to provide their students with the knowledge and experience they believe necessary for the development of positive values and attitudes. One of the tasks for environmental education must therefore be, to convince teachers of all subjects, firstly, that the development of positive values and attitudes towards the environment is worthwhile and secondly, that environmental education can be used as an aid to a more general educational process.

Commonly throughout the European Union, and beyond, secondary level teachers are teachers of single, perhaps double - usually related, subjects. For many perhaps, one of the off putting features of environmental education is the fact that proponents argue for an interdisciplinary approach (UNESCO-UNEP, 1977; Devyust & Hens, 1992). For many the thought of preparing an interdisciplinary programme of teaching and learning is daunting. This may indicate that such an approach has not yet been employed as part of their own teacher education (Bachiorri, 1995). Again the point that needs to be stressed is that an interdisciplinary programme with the environment as a focus, can be used to achieve targets and standards for individual subjects.

Table 8.1 Translating UNESCO-UNEP competencies into INSET objectives

UNESCO-UNEP	*MESAGES*
Knowledge and skills needed to identify and investigate issues.	Examine the nature of complex environmental issues and explore how these issues relate to their (individual teacher) own teaching subject.
Ability to analyse environmental issues and the associated value perspectives.	Recognise the scope for conflict between personal values and personal behaviour, in relation to both the environment and teaching practice.
Ability to evaluate, clarify and change their own value positions in the light of new information.	Reflect upon their own values in relation to the environment.

DEVELOPING MESAGES : PROCESS

Early in 1995, partial funding was secured, under a bid to DG IX of the European Commission for the development of an in-service teacher training package for environmental education for European secondary level teachers. Overall, the project was divided into four stages (See Table 8.2).

The Euro-Awareness network, which facilitated MESAGES, was organised at the time of preparing the original bid for tender. This was then elaborated during the European Seminar on Environmental Education Research, hosted by ERTCEE, in March 1995, as part of the European Nature Conservation Year. The fact that this conference was attended by individuals from all over Europe aided the process of building a network which was truly representative of the area and educational systems involved. A number of participants whose experiential backgrounds were considered relevant to the MESAGES project, (teacher educators, teachers, educational practitioners)

Table 8.2 Stages in project design

Stage	*Activity*
1	Assessment of the need for an INSET module on environmental education; Network; Proposal; Project installation (1st meeting).
2	Preparation and review of first experimental draft of module; Second network meeting; Preparation of dissemination strategies.
3	Revisions and preparation of second draft; Pilot training courses, testing and reviews.
4	Third network meeting; Collation of test results; Preparation of third draft, final commentary and amendments.

were approached by the director of ERTCEE and invited to become involved. A range of organisation types were also represented within Euro-Awareness, from university departments, non-governmental organisations to a teaching association.

Involvement within the overall project varied between participants for a number of reasons, mainly, other work pressures. Three participants left on commencement of stage two of the project. The network was strong and varied enough to adjust to this. It is entirely to their credit and resolve to raise the profile of environmental education in general, that all members of Euro-Awareness provided their knowledge, skills and time free of charge. The fact that the network and project was restricted to the European Union was determined by the call for tender. It is however, hoped that the module will be useful in other countries, particularly in Central and Eastern Europe.

June 1995, saw the completion of the first, experimental, draft of MESAGES, which was then reviewed at the second network meeting in Lisbon, Portugal. Members of the Associação Portuguesa de Educação Ambiental (ASPEA), who were also involved in Euro-Awareness, greatly assisted in the organisation of this meeting. The meeting was divided into three sections and is representative of the structure of subsequent meetings:

(1) introduction (aims and objectives of the meeting) and distribution and reading of MESAGES;

(2) general overview of the aims and objective of MESAGES itself and critical appraisal by network members;

(3) preparation for the second draft in terms of content and design.

Prior to this meeting, only ERTCEE staff had seen the draft module. Information on the aims and objectives of the module itself were deliberately withheld from Euro-Awareness members until completion of the initial reading session. This was in order to avoid readers unconsciously filling in any gaps encountered in the text.

Analysis of the first draft can be broken down under two main headings, style and content. Generally, it was felt that the initial draft took too didactic an approach and the techniques employed were not congruous with the methods supported by environmental education. It was felt that the first draft would have better served as an accompanying reader. To remedy this it was decided to increase the interactive component of the module.

Questions over the content of the module, in one sense, are on-going and can be said to have given the development of MESAGES a more organic aspect than was originally envisaged. The first version served as a discussion document and generated many of the ideas later used, either because the rejection of certain items generated particular suggestions for replacement or because they were originally absent. For example, even within the third and final version of the module, there is a section - *The Role of Evaluation in Environmental Education* - which did not previously appear. In this way the content of the final draft of MESAGES is a reflection of the consensus view of what issues should necessarily be included in an INSET programme for secondary level teachers in Europe. It is important to note that this does not mean that the final draft is an accurate reflection of the views of each individual member of the network, but of a consensus reached through discussion, debate and trial. The content of the final version is also a reflection of the fact that, as already mentioned above, in general INSET provision in Europe is approximately one day long per course.

Even with the broadest of courses, choices have to be made regarding what to include and what to omit. It was recognised by the Euro-Awareness network that these choices as represented through the publicly released version of a document like MESAGES, can have a significant influence upon the view of environmental education and what teachers think they are expected to do with that information. The slightest change in content and content sequencing can effectively alter the impression delivered. The contents pages from the three versions of MESAGES (Table 8.3) can be used to briefly illustrate this point. With this in mind, facilitators and participants of MESAGES are encouraged throughout to form and confirm their own views on environmental education using MESAGES as a stimulus.

Obviously the contents pages are not indicative of the activities and information contained within each section, but what can be seen is that in version 1, the selection of content tends to focus upon delivering and defining a number of important terms and ideas in relation to environmental education, from a theoretical point of view, for example, *Historical Evolution of Environmental Education.* The information contained within this section is not altogether absent from versions 2 & 3, it is instead presented in 'abridged' form via the section *What is Environmental Education?* In contrast to this, environmental education in the curriculum, which appears as a sub-section in version 1, is promoted in the second and third versions. By version 3, the emphasis is upon applying the concepts connected with the original terms described to secondary school practice. Version 3 is considerably more *school-based* than version 1 or even 2. On the surface this is shown by the heading for the section on *Complexity*, which by version 3, has changed to *Dealing with Complex Issues*.

Once the alterations had been made, it was agreed that network members would review the second draft of the module, using one of three methods:

- Trial, either partially or wholly, with a group of teachers or student teachers.
- Evaluation by a group of secondary level teacher trainers working within a given country.
- Evaluation by a group of secondary level teachers working within a given country.

In summary, the second draft was trailed in three countries, Scotland, Greece and Ireland. It was reviewed by teacher trainers in Belgium, France, and Germany and reviewed by teacher trainers in the Netherlands and Finland. Finally, the module was also evaluated by a number of teachers in England. In addition to this, network members who were unable to organise trials or evaluations, reviewed the second draft themselves and forwarded a commentary to ERTCEE. Although translation of the *final* draft of the module into the relevant European languages was included as part of the project plan from the earliest stages, this was found to be necessary to accommodate the evaluations. Members of the network from countries where this was required again supplied or organised the relevant skills. Due to regulations concerning the use of the French language in education institutions, even the earliest draft of the module had to be translated to enable the evaluations to take place there.

Table 8.3 Contents pages of MESAGES (three draft versions)

Version 1	*Version 2*	*Version 3*
Introduction	Introduction	Introduction
Historical evolution of EE	Glossary	What is the environment?
What is EE?	What is the environment?	What is EE?
The applied nature of EE	What is EE?	The teacher's role as environmental educator
- EE & conservation process	Complexity	Dealing with complex issues
- EE & biodiversity	Values	Developing EE in the curriculum
- EE & sustainability	Learning for action	The role of evaluation in EE
Clarifying concepts	The role of the teacher as environmental educator	Managing EE in schools
- Inter-, multi-, pluri-disciplinarity	Developing EE in the curriculum	European references
EE in schools	Managing EE in schools	Participant and facilitator evaluation sheets
- Formal & nonformal EE	European references	
- Approaches used	Evaluation sheet (for MESAGES participants)	
- Contribution on all subjects		
- EE curriculum development		
Common problems & solutions		
Sources of reference for EE		
References		

EE = environmental education

There is not enough scope within the present paper to discuss the wider implications of all these results for environmental education INSET in general, they will therefore be published at a later date. For MESAGES, however, the results indicated that what was required was to change the focus of the issues dealt with in the second draft rather than look to develop a entirely new set of topics. How this was to be undertaken was one of the central discussion points at the third meeting of the network in Bradford, November 1995. Alterations made to the module as a result of both the trails and evaluations and of this meeting constitute the final draft, subject to minor adjustments.

The process of designing MESAGES was therefore one of DO - REVIEW - REDO. Self- and peer assessment (via the network) were vital components throughout.

PRODUCT

Europe is not one homogeneous entity. This project was approached from a European aspect because the diversity of both culture and ecology in Europe - within and between countries - provides a rich source of ideas and information for environmental education. Diversity is also something that all Europeans share, in one sense it acts as a unifying force. The continued integration of Europe serves to highlight the relationship between people, their environment and their ability to survive. At the same time is serves to stress the relevance of education as a means of developing a more democratic and sustainable Europe.

Logo for MESAGES

The aims of MESAGES are to:

- present a blue-print document to assist the process of professional development for post-primary level teachers of all subjects and disciplines in environmental education throughout Europe;
- provide facilitators of post-primary INSET courses for environmental education with material, ideas and a general teaching plan for their work;
- highlight the need for teachers of all subjects to become involved in raising the environmental awareness of secondary level students
- foster the notion that environmental education can often be employed to enable students to achieve general educational targets and national curriculum standards as well as conservation and environmental targets.

There are a number of themes, or key concepts, that MESAGES hopes to promote. These themes can be employed as an instrument to guide sessions to particular conclusions if this is considered appropriate by individual module facilitators.

(1) **Conflicts:** conflicts are an outstanding feature of environmental issues and therefore environmental education. Just as motorway planners are confronted regularly with conflicting views about motorways, so too are teachers confronted with the, seeming, conflict between educational requirements and the, often, non-curricular nature of environmental education. (Although supporters of environmental education, have long since proved that in reality, this is not necessarily the case. To illustrate this, is one of the aims of MESAGES).

(2) **Role of the Teacher:** it is recognised that the teacher operates on at least three different levels - as an individual, as a classroom practitioner and as a member of a school community. Given that most teachers, although people of integrity, are not committed environmentalists, should they aim to be role-models for their students when it comes to environmental values and action?

(3) **Integration:** environmental education should not be a unique subject. Making it so would marginalise it and possibly exclude a large portion of secondary school students from receiving any environmental education. Instead environmental education should be integrated across the curriculum. Every person has a right to be involved in

decisions taken regarding their own environment. Environmental education is the most effective method of enabling people to make informed decisions on environmental issues.

(4) **Whole School Approach:** the involvement of the whole school is vital for environmental education. If the school does not show that it is concerned with its own environment, students cannot be expected to do so. Management policies, use of resources like energy, water and paper, waste disposal and school ethos are all important aspects of developing environmental education in schools.

(5) **Local and Global:** environmental education should explore both the local and global aspects of problems. Students should learn about and respect their own local environment as well as those in other countries. When they do examine issues and problems in other countries they should also be encouraged to look at their role in them, for example the destruction of Tropical Rain Forests.

(6) **Values Education:** environmental education is very often associated with science education or is considered most relevant to the natural sciences. The common assumption that underlies this is that environmental education is purely education *about* the environment, that is, there is a strong focus upon knowledge of natural processes. Knowledge *about* the social factors (*e.g.* the effects of consumer patterns or transport use) involved are as important as knowledge about chemical and physical processes (*e.g.* photosynthesis, emissions from burning fossil fuels). In this way education *through* and *for* the environment are essential features of environmental education. These features are often lacking in education about the environment via the natural sciences. Thus it is essential that environmental education should seek to encourage positive attitudes towards the environment. Further it should encourage students / teachers to act upon these values to improve their own environment. Environmental education should also provide students with an opportunity to examine the impact of their actions upon others - within Europe and beyond.

(7) **Interdependence:** this concept is central to the study of the environment. All humans are dependent on 'the environment', either directly or indirectly. Different generations - past, present and future - are also linked together via the one (changing) environment they inherit and bequeath. We do not inherit the environment of our forefathers, we inherit the effects of our forefathers upon that environment.

CONCLUSIONS

Time limitations are a considerable factor in the selection of content for INSET programmes. An inverse relationship exists between the number of topics a group such as the Euro-Awareness network, would like to encourage secondary school teachers to explore and logistical restrictions. With regard to the MESAGES project, this issue was further complicated by the different national education systems and curricula represented within the European Union. To counter these issues, MESAGES has been designed as a *generic* document. That is, it furnishes users from a variety of education systems with an opportunity to *own*, in part at least, their professional development in environmental education. It is clearly stated in the introduction that:

> *"Different outcomes will result each time this module is run, in part due to group dynamics, in part due to national requirements. It would be impossible, and perhaps undesirable, to predict what they might or should be."* (Leal Filho & O'Loan, 1996).

The depth and range of feedback received from the trails, evaluations and reviews and volume of general queries received by ERTCEE on MESAGES has indicated that there is a genuine need for environmental education INSET throughout Europe. That MESAGES is once such initiative aimed at a Europe-wide audience is certainly experimental, but so far indications are that this is a welcomed approach.

REFERENCES

Bachiorri, A. 1995. Current Trends in Environmental Education in Italy. In: Leal Filho, W., Murphy, Z. & MacDermott, F. (Eds.) *Practices in Environmental Education in Europe* Bradford: European Research and Training Centre on Environmental Education.

Council of Ministers of Education meeting within the Council of Europe 1998. *Resolutions on Environmental Education*, 24 May, 88 / C 177 / 03.

Council of Ministers of Education meeting within the Council of Europe 1992. *Conclusions on the Development of Environmental Education*, Official Journal of the European Communities, June, 92 / C 151 / 02.

Devyust, D. & Hens, L. 1992. Environmental Education at Tertiary Level. In: Leal Filho, W. & Palmer, J. (Eds.) *Key Issues in Environmental Education* Bradford: The Horton Print Group.

IUCN, UNEP, WWF 1991. *Caring for the Earth: a strategy for sustainable living*, London: Earthscan Publications.

Leal Filho, W. & O'Loan K. 1996. *Module on the Environment, Sustainability and Awareness Geared to European (secondary) Schools* Bradford: European Research and Training Centre on Environmental Education.

Oulton, C & Scott, W. 1995. The 'Environmentally Educated' Teacher: an exploration of the implications of UNESCO-UNEP's ideas for pre-service teacher education programmes. *Environmental Education Research 1* (2), pp 213-231.

UNESCO-UNEP 1976. The Belgrade Charter. *Connect I* (1), pp 1-6 Paris: UNESCO.

UNESCO-UNEP 1977. The Intergovernmental Conference on Environmental Education Tbilisi. *Connect II* (2) Paris: UNESCO.

UNESCO-UNEP 1990. Environmentally Educated Teachers: The priority of priorities? *Connect XVII* (2), pp 1-3 Paris: UNESCO.

CHAPTER 9

THE GENERATION OF BASIC ECOLOGICAL CONCEPTS - AN ALTERNATIVE APPROACH

Simon Bailey & Barbara Bajd

INTRODUCTION

Teachers have, for some considerable time now, been urged to develop 'an informed concern for the environment' in the minds of their pupils. Repeatedly, since the early sixties, bodies such as the International Union for the Conservation of Nature and Natural Resources (IUCN), the United Nations Environment Program (UNEP), the International Environmental Education Programme (IEEP) and latterly the United Nations Conference on Environment and Development (UNCED) in Brazil, have suggested that without such an 'informed concern', progress towards achieving a sustainable environment, dependent as it is upon a population capable of making informed judgements, will be seriously, if not fatally, impaired.

This historical development is well summarised by Palmer & Neal (1994). Hence a clear need arises to tease apart what is meant by an 'informed concern'; to identify the basic concepts inherent in such a definition and thence to devise strategies that will encourage the growth of these concepts, starting as soon as is feasible in the school life of the child.

The dissection process has been attempted by many groups over the years with varying degrees of clarity. Examples include IUCN (1970), Council of Europe (1971) and Carson (1971). What is clear is that any successful educational guidelines for the development of an informed concern for the environment will rest upon the foundation of a mental model of how the physical environment and the living community dependent upon it interact; on a model of the basic structure of a living community and its hierarchical relationships; of food chains and webs; of interdependence and adaptation. Put more succinctly, the foundation depends upon a base line understanding of the ecosystem concept. From such a foundation, gained perhaps from the study of small local habitats, aided by input designed to facilitate the creation of mental models at an appropriate level which will allow for better understanding of these habitats, the child should be able to approach topics such as pollution, conservation, population, renewable and non-renewable energy sources and global survival policies from an increasingly informed position.

THE NATURE OF THE PROBLEM

So how can these mental models be created? The traditional approach has been through the medium of the 'lecture' in which the relationships between levels of living organisms in the ecosystem are described, often through building the conventional 'triangle diagram' with green plants at the base and top carnivore at the apex. From this position, the role of the physical environment, of inputs into the 'triangle' and of returns from the 'triangle' to the physical environment can be added. This can then be memorised for examination purposes. *Using* the model is another matter. The extent to which a student can translate the surface-learnt model to the realities of an actual environment is debatable. Unfortunately, one cannot 'see' food chains, interdependence or population dynamics easily in the field and it is suspected that the surface learnt model is usually abandoned as being of little help in making sense of the apparently random collection of plants and animals encountered in a habitat such as wood or a pond.

We were as guilty as any in falling back on traditional methods in trying to 'get over' basic concepts, but, increasingly, became less than convinced of any real gain accruing from the approach. It was (apparently) efficient in terms of time but there was little evidence of trainee teachers being able to apply knowledge or understanding to the real situation in the field. A similar situation applied with teachers on INSET courses and with pupils in schools. In order to 'deliver' National Curriculum requirements for an English audience and similar requirements elsewhere, teachers (and their pupils) had to *understand* the basic model, should be able, at their level, to *apply* it to the real situation. An alternative method had to be utilised to bring this about.

A number of strands of thinking were, consciously or unconsciously, brought to bear upon the problem in the search for an alternative strategy. These could be grouped under two main headings:

(1) 'Lecturing' does not incorporate any *'fun'* element, for either side of the transaction. *Motivation* for the majority is often at a low level. Solution? Make the process active rather than passive. Make it entertaining. Pass the *ownership* from teacher to student.

(2) Understanding cannot be passed on from one individual to another directly. Oscar Wilde wrote 'Education is an admirable thing, but it is as well to remember from time to time that nothing worth knowing can be taught'. Understanding has to be *constructed* in the mind of the learner. This principle figures in a number of similar teaching and learning models. It is the basic tenet of the idea of Generative Learning (*e.g.* Osborne & Wittrock, 1983) and of Constructivism (*e.g.* Driver, 1985; Needham, 1987). It forms a clear distinction between surface and deep learning (Entwistle & Ramsden, 1983).

Therefore, the chosen strategy to be deployed, needed to seek out and use the ideas that individuals already had. It had to respect and value these ideas and create a climate in which they might be susceptible to modification. The strategy had to involve *thinking* on the part of the learners.

THE GENERATION OF THE STRATEGY

The idea of employing some sort of a game to bring about understanding was explored and appeared to meet many of these requirements. Board style games were discounted, largely because of personal dislike - not a valid reason for discounting the method, but the motivation in following the idea through would hardly have been uplifting. What was needed, in our opinion, was a device which was less circumscribed by rules, in which the knowledge and creativity of participants might better be harnessed. This brought in the idea of role play/drama as an ingredient. Putting the two component ideas together finally generated the overall concept from which the "Ecogame" (Bailey, 1994) developed.

Eric Berne states that *"... games are a substitute for the real living of real intimacy. Because of this they may be regarded as preliminary engagements..."* (Berne, 1964 p17). This suggests that games might be used to form ideas about the real world prior to the direct exploration of it. Hypotheses or mental models might be constructed through games for later testing in a real situation.

Karl Sigmund views games as 'thought experiments', through which play runs 'like a red thread' (Sigmund, 1993 p7). He states that *"games and play help to explore the world and teach us how to come to grips with it"* (Sigmund *op. cit.* p7). Once again, this can be interpreted as seeing games as devices for the creation of personal and group mental models, which can then be tested in the real world.

But are games 'science'? Pratt, in an article celebrating the life of Karl Popper, summarises Popper's view of the way scientists work:

> *"... a scientist is always in the situation of producing and testing tentative hypotheses in response to particular problems. The problems that are addressed have emerged from previous observations. As more testing is done, more observations accrue. The next generation of theories must explain more and so progress in science corresponds to the evolution of more powerful explanatory theories."* (Pratt, 1995).

Popper's perspective is mirrored in the view of learning in science put forward in the non-statutory guidance issued in connection with the original (1989) National Curriculum document for science and not superseded with the issue of later versions (1991 & 1995): *"In their early experiences of the world, pupils develop ideas which enable them to make sense of the things that happen around them. They bring these informal ideas into the classroom and the aim of science education is to give pupils more explanatory power so that their ideas can become useful concepts."* (NCC, 1989 para 6.2).

Thus, ideally, games devised for developing ecological understanding ought to take these views on board. They should utilise the individual and collective initial views of the pupils; encourage the development of hypotheses from data collected and allow for these to be subjected to testing in the real world and should permit pupils at differing stages in development to refine their hypotheses so as to expand their explanatory power.

The central idea of the game is that pupils take on the role of a character within an ecosystem and then compete, as the organism competes, for survival. They feed and are fed upon, using cards to represent the living material of their bodies and these cards are passed backwards and forwards between players, with the names of successive owners of the material being recorded, so as to provide the data which could then be used to develop the concept of food chains. This succession in 'ownership' fosters the idea that they need other living things to survive, that is the idea of interdependence. The numbers of cards that individuals hold moves them towards the idea of 'getting bigger' (growth) or of there being 'more of them' (reproduction). It can also be developed towards the concept of population dynamics.

The game has a fairly tight structure, at least in the early stages and the role of the game leader is crucial in leading the pupils progressively towards the learning objectives. However, the tight structure is not over-restrictive for the players as there are very few 'rules' (indeed most of the rules are made up by the pupils as the game progresses). The pupils are only rarely 'told' anything by the leader, careful questioning leads towards the pupils coming to a consensus from the basis of their own existing knowledge. So, to a certain extent, the game is constructivist in its strategies in that it builds a model from what the pupils already know.

THE STRATEGY OUTLINED

The several stages in the game lead towards an understanding of food chains and webs, interdependence, decay and materials cycling.

These stages are described in detail elsewhere (Bailey, 1994). The game starts with a statement that the objective is to find out how (or if) living things 'fit together' in the real world and that this exploration is to be undertaken through using a 'sort of play'. The players are given roles, often with a humorous slant (*e.g.* Smelly the Fungus, Bigmouth the Frog) which tend to increase motivation, at least with younger pupils. The roles are chosen to include *all* levels within an ecosystem (*e.g.* primary producer, herbivore, lower and upper carnivore, decomposer) but these are *not* explained to players at this stage. The first task of the players is to decide if they represent, collectively, a particular place, that is if particular groups of plants and animals are to be found in particular environments.

Once this has been achieved, the process of ascertaining how players might acquire materials to power growth and reproduction is explored. This is apparently done at random, but in reality must be carefully structured so that it builds from the green plants, through herbivore and carnivore to decomposer. As a consequence of this, all players establish a rationale as to how they might gain cards (to represent living material) either from the group leader (who represents the physical environment with its supply of air, water and sunlight) or from other players who have already established how they might obtain cards. At the same time, the principle of card exchange, with the routine of writing the owner's name on the cards as soon as they are obtained, is established and reinforced several times as the 'random' progression through the hierarchy of organisms develops.

Once all players have gained cards through discussion with the game leader and among themselves, data cards can be issued to each player, informing them of 'who they can eat and who can eat them'. This procedure avoids over-enthusiastic oak trees eating badgers in an attempt to 'get as many cards as they can'. The system has worked well, but could be replaced by specially constructed data files on a simple database.

The most active part of the game, a free exchange of cards to represent the dynamics of the habitat, then commences. However, this must be prefaced by a reminder of the rules of card exchange (departure from these rules will invalidate the game) and a strategy for solving problems that will arise during the active play. It is from the *group* resolution of problems that much of the real learning appears to arise.

In order to achieve rapid cycling of 'living material' through the system, the cards (three) given initially to each player are clearly different to those given later to the producer organisms (provided they can justify this gift) by the game leader to represent their continued growth after they have been 'eaten'.

The active stage of the game finishes at the point when the leader can see evidence of food chains on at least some of the cards held by the top carnivores. The players can now be invited to analyse the cards that they and the other participants hold, given careful structuring, and questioning by the leader.

ANALYSIS OF THE GAME PLAY

Taking a good example (with at least three to four 'ownerships' involved) from a top carnivore and working back from this organism through previous owners to the green plant level, concrete evidence for the concept of food chains can be built up in front of the players without any technical terms being used. The players *see* it, they are not told it. Having themselves arrived at the concept of and term for food chains, they can speculate as to how their example could be extended at either end, thus bringing in the ideas of connection to the physical environment (the game leader) at one end and the idea of decomposer organisms 'eating' the top carnivore at the other. This can then be developed to include waste materials produced at this (and other) level(s).

The realisation that few of the completed cards are identical can now be used to extend the idea of food chains to food webs. Indeed, all the cards produced during the game, or a sample of them, can be used to produce a food web for 'their' habitat. When this has been done the similarity between the game-generated version and typical textbook examples is remarkable.

The role of the decomposer organism can now be extended to a consideration of the concept of nutrient cycling, using the existing knowledge of the players as evidence. If, for example, Wriggly the worm realises that he/she can feed either on a dead fox or, more likely, on the waste from a fox, she/he will probably accept that the waste of the worm could also be used as 'food' by other living things. As worm casts are easily seen by children as 'soil', the connection back to the producer organisms can readily be made.

If the 'food chain' already described has concrete form (as a line of players with linked arms), the extension of the food chain concept at either end is simple to achieve by the addition of other players. Once the connection back to the producer organisms from the decomposer is floated as an idea, concrete display of this can be gained through linking the two ends of the chain to form a circle. Hence the concept of nutrient cycling can be given a physical form. Equally, the food web could be displayed, but the complexity of linkages generally presents physical problems for pupils within a very short time. This demonstration of the complexity of ecosystems can prove a valuable learning opportunity.

The elements of the game as described thus far should have established some basic patterns within a living community. These patterns will have become apparent as a result of the ideas and actions of the players; *they* will have created the evidence to suggest them. Thus, any generalisations about the way living things interact with each other and with the physical environment will take the form of self-constructed meaning, rather than received 'wisdom'.

DEVELOPING ENVIRONMENTAL EDUCATION ASPECTS

As yet, the game is flawed from the point of view of developing environmental awareness in that it excludes the Human impact. But a state of readiness should now exist to adapt any models created to include this factor. The level to which this is developed will depend upon the age and experience of the players. It could be, with young children, that they be invited to speculate on the 'previous history' of their lunch! The idea that their lettuce sandwiches were once 'worm or cow poos' (common 7-11 year old terminology for faeces) tends to stay in their minds!

They should be ready to consider what might happen if 'we killed all the foxes' or 'a disease killed all the rabbits'. Aspects of conservation can develop from such small starting points. It is easy to dismiss such a statement as superficial but it is perhaps worth noting that the development would be based upon *understanding* from a self-created mental model of the basic ecosystem, not upon a transmission-led introduction to the topic of conservation.

If the game succeeds in creating mental models of food chains, interdependence, material cycling *etc*. (and trials of the game strongly suggest that it does), then these will only be at the level of tentative hypotheses. These ideas need to be tested by exploring real habitats, by looking for evidence in them to support or refute the models created through the game. Indeed, learning potential is seen in pupils creating, and then playing, their own version of the game, based upon the flora and fauna of a habitat that they have studied. They could even explore the possibilities of creating a game for an unseen habitat, such as the rain forest.

MODIFYING FOR DIFFERING AGE RANGES/LEVELS OF EXPERIENCE

The basic game, as described in outline above, has been developed with pupils aged 7-14 in mind. It has been tested with most of the year groups within this band, in a variety of settings and has, on all occasions been adjudged successful as a means of bringing about learning. It has also been trialed extensively throughout its development with groups of teachers training for the age band and with teachers on in-service courses, with equal success. However, the game would need modification if it were to be used with either younger pupils or with older, more experienced players. The basic learning targets of the game remain unchanged, whatever the age and experience of the players. The complexity of the teaching strategies and the depth to which the concepts raised are pursued will form the major modification areas.

A version of the game, suitable for 5-7 year old children, is currently under construction (Whitney, 1995). Modifications involve a reduction of the 'cast list' from 26 to 4 characters, one each from producer, herbivore, lower and upper carnivore levels. There will be a much greater emphasis placed upon prior art work (*e.g.* the making of character masks), upon the use of music and movement to encourage ideas of growth and feeding, upon a carefully structured 'script' for the leader and upon the growth of the food chain concept as the game progresses, rather than on building this in at the end of the game. Sub-concepts, such as decay and the role of decomposers, will probably not feature in the basic game, although provision may well be made for them to be brought in afterwards. With young pupils, there may be a stronger case for repeated playings of the game, perhaps at increasing levels of complexity, than would be the case with older pupils.

Modifications for students above the age of 14 relate particularly to the concept of energy flow in ecosystems as well as to the greater spread of concepts that can be brought in at the analysis stage, based upon the mental models created.

Essentially, the energy aspect is developed from adopting a dual card system in the basic game play. Different coloured cards are used to represent raw materials (water and air at the simplest level, developing to bring in the significant components of air and mineral nutrients at higher levels) on the one hand and energy (either as light energy input or as heat energy output) on the other hand. The bringing together of pairs of these cards at the producer level during the process of photosynthesis establishes the idea of energy becoming 'trapped' in the monosaccharide molecule produced. Hence the role of respiration in 'releasing' this energy to power other living processes in the producer or the role of 'passed on' monosaccharide or its derivatives in powering the living processes of the organism that eats the producer can be perceived from concrete 'evidence'.

Equally, at the decomposer level, the 'taking apart' of the complex molecules of waste material so as to produce inorganic material will 'liberate' trapped energy into the physical environment. The change in energy form from light to heat can be explored and discussed.

The level to which the energy aspects of the game are developed will, of course, vary according to the age and experience of the players. The inclusion of semi-quantitative aspects of energy flow for biology undergraduates might merit serious consideration. This could be achieved through arbitrary reduction of the energy passed on at each exchange in the game, achieved perhaps by the cutting of cards with the resultant return of fragments of energy cards to a 'background heat' box. The raw materials cards could be subjected to a similar process of sophistication in processing.

The development of associated sub-concepts becomes even more natural when dealing with students who have a greater personal database at the time of playing the game. The ideas of pyramids of numbers/biomass/energy can be illustrated, adaptation can be introduced through incorporating a device that allows the 'eaten' to reject the advances of an 'eater' at a projected card exchange through stating a (real) mechanism by which they might avoid being consumed. For example, and at a simple level, the hedgehog could rebuff the fox by 'rolling into a ball'. The bramble could reject the rabbit by 'being thorny'.

The effects of day, night and season could be brought in along similar lines. The thrush could avoid card exchange with the owl by claiming that one is diurnal, the other nocturnal. The owl could respond at a later exchange by stating a dusk encounter! The hazel could reject the advances of a squirrel by claiming that it was summer and the nuts were not yet ripe. The game environment encourages the players here to use their creativity and imagination, balanced by an increasing knowledge of the realities of plant and animal life, to develop their ideas on concepts such as adaptation.

The idea of and probable effects of species introduction could be incorporated. The effects of Humans upon the ecosystem have already been touched upon, as well as the introduction of the myxoma virus; these could be considerably expanded with older, more experienced students.

MODIFYING THE GAME FOR DIFFERENT CULTURES

Some suggestions have been made as to the modification of the game for differing ages whilst retaining the principal learning targets. Equally, it is considered that the basic game has potential for modification to suit differing cultural settings. Some experience has been gained in translating the game (both at an early stage and at later stages of development) to meet the needs of INSET groups in Slovenia. Essentially, all that was required was to modify the 'cast list' so that the examples of the various hierarchical levels in the ecosystem were represented by locally present species. The modification was achieved in a fairly short time, aided by the fact that the team responsible had a background in biology and an understanding of binomial nomenclature! It was complicated when the game was at a very early stage in its development and only one member had any prior understanding of the nature and purpose of the game at the time. Once these items had been clarified and published, translation to the Slovenian setting proved a relatively simple operation.

Providing the game for a different culture and in a different language was aided by the use of translators. This worked well, largely owing to the fact that the translators were trained biologists with good ecological backgrounds. They were thus able to respond positively to queries from players without reference to the non-Slovenian speaking leader. It is considered that there would be few problems in adapting the basic game to other cultural settings.

Further experience was gained during trial work in using translators. On two occasions, the game was played with classes of 8-10 year olds in a school with a majority (80%) of Asian children, many of whom were not fluent in English. A bilingual classroom assistant translated where necessary, amplifying and explaining the leader's words to pupils or bringing problems to his attention (Bailey, 1995). It is worth noting that she grasped the intention of the game in a very short time-span, indicative perhaps of the speed with which trained (or even untrained) teachers could make use of the strategy.

THE ADVANTAGES OF THE STRATEGY OVER TRADITIONAL METHODS

What then are the advantages, if any, of utilising a strategy that employs role play/drama techniques? Trials in a variety of schools have given several pointers to advantage gained. It should be noted that, although these advantages were perceived in all the trial schools, without exception and were also envisaged by teachers on INSET courses as potential advantages for the children they taught, they have not yet been quantified in any way. Research is currently being planned on the comparative gain in understanding arising from 'traditional' approaches and a strategy involving the use of the "Ecogame" as a starter activity in the teaching sequence (Watson, 1995).

There is clearly a high level of pupil motivation and interest when the game is played. The approach is perceived as 'different'; there is considerable pupil activity; the pupils are required to *think* during the game, they have to solve real problems. There is a slight competitive element in that each player is trying to 'get the most cards'. There is very little requirement to write during the game play, other than to attach their role names to the cards.

There is strong evidence that the game works well in mixed-ability situations and that it, almost automatically, brings about differentiation of task for the full ability range, without further modification. It has proved taxing to the highest ability levels while bringing about high time on task and clear learning gain for the considerable proportion of pupils who, for one reason or another, have been labelled as possessing 'special needs'. Occasionally, specific strategies have been adopted to cater for some of these children. The most common modification has been the pairing of a special needs child (usually with a writing handicap) with another child in playing a role. This has brought about learning outcomes for these 'special' children that have surprised their teachers. Generally, such pupils derive enjoyment from playing the game; their time on task is high and their concept acquisition appears considerable.

CONCLUSION

No claim is made that the strategy outlined, is the 'best' way to develop ecological understanding. However, it has at all times been well received by pupils and their teachers and seems capable of being modified to suit all ability and age levels. At the lower levels of education, certainly below age 11, an extensive ecological knowledge is not required on the part of the teacher, so is perceived as non-threatening by the vast majority. This has certainly been the case in the fairly large number of primary and secondary teachers involved in the trials.

There is always a case to be made for variety in teaching styles; not least in the teaching of science which has, at times, suffered a 'bad press'. Non-quantified trial results suggest considerable conceptual gain, high levels of pupil motivation and time on task whatever the age or ability of the group. Can one say the same of traditional didactic methods?

REFERENCES

Bailey, S. 1995. Developing ecological understanding through drama: The Ecogame. *Primary Science Review 38*, pp 11-13.

Bailey, S. 1994. The Ecogame Resources for Learning for BNFL/ Manchester Metropolitan University.

Berne, E. 1964. *The games people play* Penguin.

Carson, S. McB. (ed.) 1971. *Environmental Studies: The construction of an A level syllabus*. National Federation for Educational Research in England and Wales.

Council of Europe 1971. Committee of Ministers. Appendix to Resolution (71) 14 Strasbourg: Council of Europe.

Driver, R., Guesne, E. & Tiberghien, A. 1985. *Children's ideas in science* Milton Keynes: Open University Press.

Entwistle, N. & Ramsden, P. 1983. *Understanding student learning* London: Croom Helm

International Union for the Conservation of Nature and Natural Resources 1970. *Final Report of the International Working Meeting on Environmental Education in the School Curriculum* Paris: UNESCO.

National Curriculum Council 1989. *Science. Non-Statutory Guidance* York: NCC.

Needham, R. 1987. *Teaching strategies for developing understanding in science* Leeds: Centre for Studies in Science and Mathematics Education, University of Leeds.

Osborne, R. J. & Wittrock, M. C. 1983. Learning science: A generative process. *Science Education 67* (4), pp 489-508.

Palmer, J. & Neal, P. 1994. *The Handbook of Environmental Education* London: Routledge.

Pratt, A. 1995. Karl Popper and science. *New Zealand Institute of Chemistry Publication No 59*, pp 23-31.

Sigmund, K. 1993. *Games of life - Explorations in ecology, evolution and behaviour* Oxford: Oxford University Press.

Watson, R. 1995. Personal communication.

Whitney, C. 1995. Personal communication.

INDEX